Essential Data Analytics, Data Science, and AI

A Practical Guide for a Data-Driven World

Maxine Attobrah

Apress®

Essential Data Analytics, Data Science, and AI: A Practical Guide for a Data-Driven World

Maxine Attobrah
New York, NY, USA

ISBN-13 (pbk): 979-8-8688-1069-5
ISBN-13 (electronic): 979-8-8688-1070-1
https://doi.org/10.1007/979-8-8688-1070-1

Managing Director, Apress Media LLC: Welmoed Spahr
Acquisitions Editor: Celestin Suresh John
Development Editor: Laura Berendson
Coordinating Editor: Gryffin Winkler

Cover image designed by Image by kjpargeter on Freepik

Distributed to the book trade worldwide by Springer Science+Business Media New York, 233 Spring Street, 6th Floor, New York, NY 10013. Phone 1-800-SPRINGER, fax (201) 348-4505, e-mail orders-ny@springer-sbm.com, or visit www.springeronline.com. Apress Media, LLC is a California LLC and the sole member (owner) is Springer Science + Business Media Finance Inc (SSBM Finance Inc). SSBM Finance Inc is a **Delaware** corporation.

For information on translations, please e-mail booktranslations@springernature.com; for reprint, paperback, or audio rights, please e-mail bookpermissions@springernature.com.

Apress titles may be purchased in bulk for academic, corporate, or promotional use. eBook versions and licenses are also available for most titles. For more information, reference our Print and eBook Bulk Sales web page at http://www.apress.com/bulk-sales.

Any source code or other supplementary material referenced by the author in this book can be found here: https://www.apress.com/gp/services/source-code.

If disposing of this product, please recycle the paper

To my family, friends, and teachers, for your endless love and support,

To my dad, who believed in me before I believed in myself,

To my mom for always praying for me,

To God, for guiding my steps and granting me the strength to persevere,

And to all the dreamers, may your journey lead you to wonderful places.

Table of Contents

About the Author

Maxine Attobrah holds a bachelor's degree in Electrical Engineering from the University of Massachusetts – Amherst. Maxine's career began as an Electronic Flight Controls Engineer at a leading global security, defense, and aerospace contractor company, where she was responsible for developing and testing control system software to enhance helicopter piloting. Subsequently, Maxine pursued further education, earning master's degrees in Electrical and Computer Engineering and Engineering & Technology Innovation Management from Carnegie Mellon University. Maxine started her career after graduating at a major global consulting firm as a Data Scientist and has since transitioned to the role of an AI/ML Engineer. Currently, she serves as a Lead AI/ML Engineer at this firm.

This book was prepared by the author in her personal capacity. The views and opinions expressed in this book are those of the author and do not necessarily reflect the official policy, opinion, or position of their present or past employers.

About the Technical Reviewers

Divit Gupta is a highly experienced IT professional with two decades of industry expertise, renowned for driving strategic architecture initiatives and providing leadership in multi-pillar sales cycles on a global scale. He leads technical partnerships, shapes team visions, and champions innovative strategic endeavors.

In addition to his corporate achievements, Divit is a prominent figure in the world of technology podcasts. As the host of popular shows such as "Tech Talk with Divit," "Live Labs with Divit," and "Cloud Bites with Divit," he highlights significant technological initiatives and provides insightful leadership. In 2022–2023, he served as Oracle TV's correspondent for Cloud World, further showcasing his commitment to knowledge sharing. Divit's dedication to sharing knowledge extends to international conference talks, technical blogs, and authoring multiple books on emerging technologies. He is a recognized expert who presented on Oracle Database technology at Oracle Cloud World FY 2023. With over 40 certifications from Microsoft, Oracle, AWS, and Databricks, he continues to remain at the forefront of the technology landscape.

Meet **Prashanth Lakshmi Narayana Chaitanya Josyula**, a dynamic force in the tech world whose journey is marked by an unyielding passion for innovation and an extraordinary depth of expertise in both technical literature and software engineering. As a Principal Member of Technical Staff (PMTS) at Salesforce, Prashanth doesn't just meet expectations – he consistently exceeds them, pushing the boundaries of what's possible in technology.

With over 16 years of robust experience in the IT industry, Prashanth has mastered a multitude of programming languages and technologies, establishing himself as a true polyglot programmer. His proficiency spans across Java, Python, Scala, Kotlin, JavaScript, TypeScript, Shell Scripting, SQL, and an array of open source solutions. Since beginning his professional journey in 2008, he has delved into various domains, each time leaving a mark of excellence.

In the realm of Java/Java EE and Spring, Prashanth has been instrumental in designing and building resilient, scalable backend systems that power critical applications across industries. His deep understanding of these technologies ensures robust and high-performance solutions tailored to meet complex business needs.

Prashanth's expertise in UI technologies is equally impressive. He has crafted intuitive, responsive user interfaces using frameworks like ExtJS, jQuery, Dojo, Angular, and React. His commitment to creating seamless user experiences shines through in every project, bridging the gap between complex backend processes and user-friendly front-end interfaces.

Venturing into big data, Prashanth has leveraged platforms like Hadoop, Spark, Hive, Oozie, and Pig to transform massive datasets into valuable insights, driving strategic decisions and innovations. His ability to harness the power of big data showcases his analytical mindset and his knack for tackling large-scale data challenges.

In the field of microservices and infrastructure, Prashanth has been a pioneer in engineering robust and scalable solutions with cutting-edge tools like Kubernetes, Helm, Terraform, and Spinnaker. His contributions to open source projects reflect his commitment to collaborative innovation and continuous improvement.

Moreover, Prashanth is at the forefront of AI and machine learning, exploring and advancing the capabilities of these transformative technologies. His work in this area is characterized by a fearless approach to experimentation and a relentless pursuit of knowledge.

Each day for Prashanth is an exciting adventure, filled with opportunities to learn, innovate, and lead. His career is a testament to his dedication to advancing technology, not just for the sake of progress but to truly make a difference. With his unparalleled skills and a visionary mindset, Prashanth continues to inspire peers and push the envelope of technological possibility.

Foreword

With breakthrough innovations in artificial intelligence and the ensuing hype, everybody is now talking about AI – and that's whether they know what they are talking about or not. And the topic is important as the ability to understand and leverage data is more critical than ever. Yet, with so much technology, jargon, and code, exploring AI can be intimidating and overwhelming. This book aims to break down those barriers, offering an approachable and concise guide to the world of data science and machine learning.

This book takes readers step by step through the entire data science process – from obtaining, preparing, and processing data to performing exploratory data analysis and training and evaluating models. Crucially, it doesn't stop at model training. The book also covers the essential next steps: deploying and monitoring models, improving them through experimentation with live data, and addressing key considerations like security, trust, and how to make models work effectively in practical, real-world scenarios.

Focused on what truly matters, the book avoids bogging readers down with excessive code while providing numerous examples and illustrations. Again and again, it discusses multiple practical use cases and the decision-making involved.

For anyone eager to cut through the AI hype and gain a real understanding of data science and machine learning, this book serves as a concise, practical, and accessible companion – empowering you to confidently navigate the fast-changing world of AI.

Christian Kästner, Associate Professor and Director of the Software Engineering PhD Program

Carnegie Mellon University, School of Computer Science

CHAPTER 1

Introduction

Have you ever wondered how Netflix knows what show you will binge next? Or how Siri, Alexa, and ChatGPT always have an answer to your weirdest questions? What about when you wake up in the morning and order something online or check your newsfeed on your favorite social media app? Have you ever wondered how all these personalized experiences came to be?

The three magicians behind the curtain are data analytics, data science, and artificial intelligence. The key behind all these three is data! Every click, every swipe, and just about every action leaves a trail that gives you a unique experience. This book will help guide you to understand these three concepts and how data is used to shape the world we live in.

Data Analytics

Imagine you are a detective, and your task is to solve mysteries. However, instead of relying on a magnifying glass like Daphne from Scooby-Doo, you have a more powerful tool – data! Now, you can perform data analytics. Data analytics turns raw data into insightful stories. Just like a detective looks for clues, data analysts sift through data to find patterns. With the power of data analytics, questions become answers. They can answer questions like "Why did sales drop in July?", "Which song will be a hit next summer?", or "Why do we keep running out of vanilla ice cream so quickly on Saturdays?".

M. Attobrah, *Essential Data Analytics, Data Science, and AI*,
https://doi.org/10.1007/979-8-8688-1070-1_1

Data Science

Data analysts help find patterns, but now that you have uncovered the patterns, what is next? Data scientists help predict the future. They have a crystal ball that predicts the future by understanding the past and the present. They are the bridge between data patterns and meaningful actions. When your favorite shopping app suggested those pairs of shoes or your email service provider removed that spam email from your primary inbox before you could see it, that was all the work of a data scientist. Imagine predicting the next big fashion trend or forecasting stock prices. You must wonder how they must get this done.

Have you heard of Python? No, not the snake. Have you heard of R? No, we are not going to sing our ABCs. Python and R are programming languages. Data scientists have multiple programming languages within their toolbox. These tools make other tools, like neural networks and random forests, which are used to make predictions. Don't worry; by the end of this book, you will better understand these concepts and how to use them.

Artificial Intelligence

Neural networks and random forests are part of artificial intelligence (AI). Data scientists use AI to make predictions. AI teaches machines to think and learn. It's like having a device as a virtual assistant. It can do more than forecast stocks or predict the next fashion trend. AI is used in software that paints like Picasso. It is in some software that composes music. It's in automated cars and can be used to create a chatbot that answers your queries. There are many subdomains of AI.

Nonetheless, with great power comes great responsibility. Some questions need to be answered: Does my software need AI? Should machines make decisions for us? What if it goes rogue? The world of AI is fascinating, but there are many issues to consider as well.

We will explore real-world applications of AI across various industries, illustrating both current and potential use cases. Additionally, societal impacts and ethical considerations will be thoroughly discussed in later chapters of this book.

The Dream Team

Data analytics, data scientists, and AI together make a super team. Together, they shape how we live, work, play, and dream. We now live in a world where your fridge can let you know when your milk is running low, and you can get a personalized playlist generated based on your mood.

The world is data-driven, and it's time to embrace it. Whether you're interested in becoming a data detective, a data science sorcerer, or an AI aficionado, there is a place for everyone on this stage. Remember, as this chapter comes to an end, every click, every swipe, and just about every action you make leads us to another step into the future. Embrace it, enjoy it, and watch the magic unfold.

What to Expect?

We will cover various topics in this book to strengthen your understanding of how all these things work individually and together. We will start by discussing how to obtain data. Data is your most essential tool, whether what you are building is just for practice or real-world situations. Some people pay to get data, but there are ways to get this data for free. Sometimes, you may encounter situations where you must create the data yourself. This is called synthetic data creation. I will show you ways you can do that as well. There are pros and cons to all these methods.

After you obtain data, you will need to go through a process called ETL. This is an essential process when dealing with data. ETL stands for Extract, Transform, and Load. There are commercial tools that can do this for you. However, you can also create this pipeline on your own using tools I have mentioned earlier, like Python or R. Once you have loaded the data to whatever destination you would like, you should explore and understand the data you now have in your possession. People in the data world call this EDA, which stands for exploratory data analysis. There are commercial tools that can help you with this. For example, you may have heard of Tableau! This is a popular tool but can be expensive for some people to get access to, so there are free ways to get this done.

Next comes what some people may think is the fun part after you have loaded your data and cleaned it! It is essential to feed your model clean data. Algorithms are only as good as the data used to train them. You can use different algorithms to create models depending on the tasks you are trying to accomplish. After training the model, you need to evaluate it. Do not deploy a model without testing it does what you want it to do. There are various techniques for assessing a model. Not all evaluation techniques work or are necessary for the same model.

As you can imagine from some of the examples I mentioned earlier, AI can live in many different places. Once you have decided you are ready to deploy your model, it is time to determine where it will live. It can be placed in the cloud or on a device like your mobile phone, fridge, or even your car! Depending on where it is placed will determine how fast or slow it takes to make a prediction or a decision.

After giving your model a home, that is not the end. It is never the end! You must continue to nurture your model and keep it safe from bad actors. We will discuss some different adversarial attacks to look out for. We will discuss how to keep your model up to date using something called telemetry.

It is important to note that not all tasks require the use of AI. Depending on the tasks, the use of AI can be overkill. You should take into consideration things like cost, size, and other stuff I will discuss in later chapters.

Finally, there is an understandable fear of the power of AI. Unfortunately, I do not believe AI is going anywhere, so you should embrace it and find ways to work with it instead of against it. Of course, laws should be in place to limit how far we can take it so we do not all end up zombies submissive to the AI gods. However, many things have changed and will continue to change. It is not as scary as people think. No. The nerds will not kill us all. Let's learn more about it and see how we can use it to help us grow in our daily lives to be more efficient with our time and routine.

It is important to note that the field of data analytics, data science, and AI is rapidly evolving. While there may be new developments that are not covered in this book, the foundational principles discussed will remain relevant. These core concepts will continue to serve as a valuable guide, even as technologies and methods evolve.

CHAPTER 2

Obtaining Data

Data is one of the most essential tools in data analytics, data science, and artificial intelligence. There are different types of ways data can be presented to you. Most data are messy and need to be cleaned before being used.

This chapter will delve into the multifaceted world of data, unraveling its various types and exploring the many avenues it can be acquired. We will navigate the intricacies of extracting data from the Web, acquaint ourselves with the treasure troves of public datasets, and demystify application programming interfaces (APIs). We will also touch upon the traditional yet vital method of collecting data through surveys and questionnaires. Finally, we will discuss ethics and responsibilities in data collection.

Understanding the Different Types of Data

Let's discuss the key distinctions between different types of data to effectively guide the process of gathering and analyzing data in your projects.

Structured Data

Structured data is organized data that is easily readable by both humans and machines. This type of data is usually found in tools you are used to seeing, like spreadsheets and relational databases. The data is assigned a

M. Attobrah, *Essential Data Analytics, Data Science, and AI*,
https://doi.org/10.1007/979-8-8688-1070-1_2

column, and each row refers to an instance of that data. This type of data is usually called tabular. Tabular data is organized in a table with columns and rows.

Structured data is beneficial because various tools and techniques can easily search and analyze it. It is also easier for any person to read without any special tools. This type of data is ideal for business intelligence, data analytics, and artificial intelligence applications.

Unstructured Data

Unstructured data is data that is not in a specific format or structure. Most data in the real world are in this form. Unstructured data include invoices, images, audio, video, and email. Some examples of this kind of data can be seen in Figure 2-1.

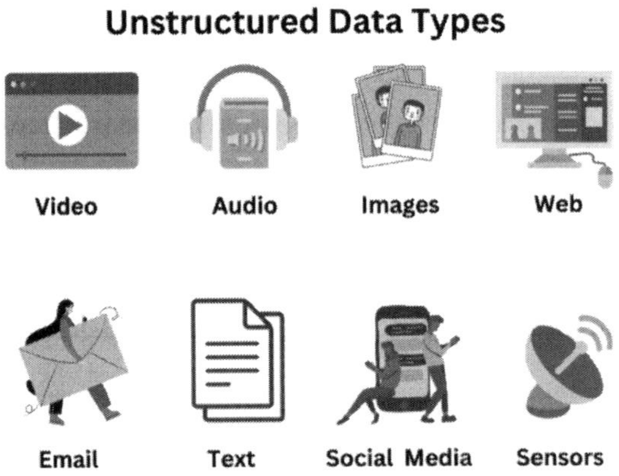

Figure 2-1. *Images of examples of unstructured data*

The difference between structured and unstructured data is the type of analysis that can be done with it. Structured data is in a rigid format that ensures consistency. Unstructured data, on the other hand, is not uniform. Figure 2-2 shows a visual of the difference between the two types of data.

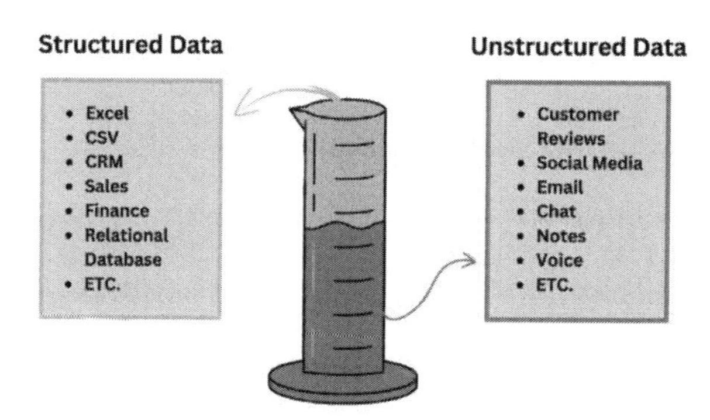

***Figure 2-2.** Image of examples of structured data vs. unstructured data*

Quantitative Data

Quantitative data is anything with numerical properties. This data can be counted or measured. You can answer questions like how many, how much, or how often with this type of data.

Qualitative Data

Qualitative data is information that **cannot** be counted, measured, or expressed using numbers. You can answer questions like why, how, or what happened.

In Figure 2-3, I provide some ways to think of qualitative vs. quantitative data.

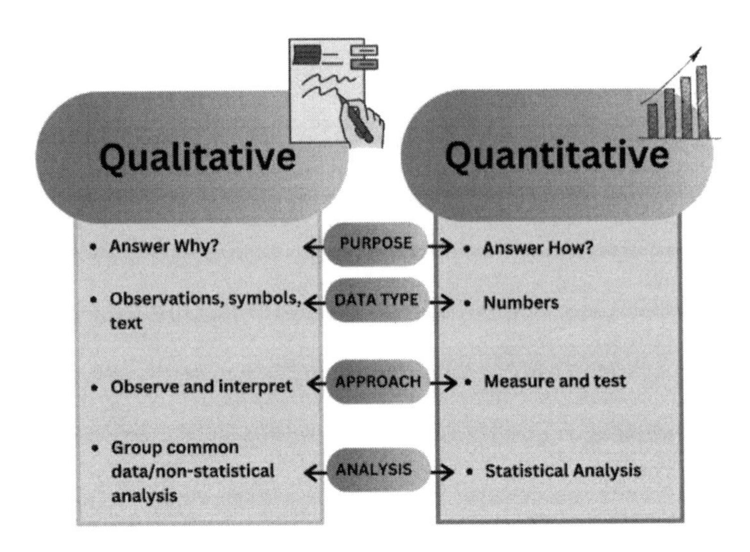

Figure 2-3. *Image of examples of how to think of qualitative data vs. quantitative data*

Time Series Data

Time series data is data collected in different time periods. This type of data helps with understanding trends and seasonality. Trends show the directions of the data. It reveals that data is increasing, declining, or remaining the same over a specific period. For example, e-commerce trends show a growth or decline in sales in the last year, which can help you adjust for the next fiscal year. Seasonality shows patterns that occur regularly – for example, a spike in vanilla ice cream sales on Saturdays.

Cross-Sectional Data

Cross-sectional data is data collected in one single moment in time. This type of data helps understand the relationship between items at a specific time.

Spatial Data

Spatial data or geospatial data is data that references a particular location on Earth. For example, you can see the longitude and latitude in each row of your data. This data is commonly analyzed using a Geographic Information System (GIS). A typical software tool used is ArcGIS.

Ways to Obtain Data

Let's explore various methods for gathering data, from traditional surveys, web scraping, and even accessing popular websites that provide datasets that are downloadable, to fuel your analysis.

Web Scraping: Extracting Data from the Web

Simply put, web scraping is the process of extracting content from a website. You could copy/paste data from a website. However, depending on how much data you want to retrieve, that process can take a lot of work. By using a tool such as a programming language, you can reduce your time from hours to seconds while you grab yourself another cup of coffee to get through the workday. It is important not to infringe on a person/company's copyright.

API: Application Programming Interface

APIs are a way for two or more software to communicate without knowing all the details within the application. It acts as an intermediary layer that processes data transfers. This saves developers time and encourages collaboration between different applications. APIs come with documentation that can be considered a contract representing an agreement between two parties. It will tell you if Party A sends a remote request structured in a predefined way, then Party A can expect a particular output from Party B. This allows companies to share their data and functionalities with third-party developers and business partners. It is like ordering food from a specific menu at a restaurant. When you request a particular item, you expect to receive it without knowing how it was created.

Questionnaires and Surveys

Questionnaires are written sets of questions used to ask individuals or a group. Surveys are a research method that uses questionnaires to help people understand the bigger picture by evaluating the data and drawing conclusions. You can use tools like Google Forms and Typeform to create surveys.

Popular Sites and Datasets

I have a few favorite sites I will share with you that you can use to access data. These sites can serve as valuable resources for obtaining the data you need. In the vast landscape of data analytics, data science, and artificial intelligence, the availability of datasets is very important for meaningful analysis and model development. The following websites provide access to a diverse range of datasets, catering to various domains and interests.

Kaggle

Kaggle is an online platform for data science and machine learning enthusiasts. It was founded by Anthony Goldbloom and Jeremy Howard in 2010. It was acquired by Google in 2017. Users on this platform can collaborate with other users, upload datasets, and compete against one another to solve data science challenges. Kaggle is a great place to get free datasets and practice your skills. People worldwide with varying backgrounds on this platform learn and challenge themselves individually and together.

Throughout this chapter, I have spoken a lot about tabular data. However, on platforms like Kaggle, you can find various types of data – for example, images.

The Facial Expression Recognition 2013 (FER2013) dataset has 35,887 grayscale images that are 48x48. There are seven emotions: anger, disgust, fear, happiness, sadness, surprise, and neutral. This can be used in artificial intelligence to perform tasks like emotion detection. Figure 2-4 shows some examples of the types of images within this dataset.

Figure 2-4. *Random images of facial expressions in FER2013*

Hugging Face

Hugging Face is another online platform for data science and machine learning enthusiasts. One of the significant differences between Kaggle and Hugging Face is that there are no competitions. It provides members with a way to find datasets as well as a way for members to discover, train, and deploy models. It was founded by Clement Delangue, Julien Chaumond, and Thomas Wolf in 2016.

EXAMPLE WITH CODE

The following sites I will mention will come with examples using Python. To follow along, you can use software applications like Jupyter Notebooks, Google Colab, or even Visual Studio Code, which are all free.

Jupyter Notebook

Jupyter Notebook is an interactive web application that you can use to share code, equations, and visualizations. It supports various languages, including R, Python, and Scala. The notebooks are typically saved in .ipynb formats but can be converted to formats like .py. Data scientists use it for data exploration, machine learning modeling, and experimentation.

Google Colab

Google Colab is a free Jupyter Notebook environment hosted in the cloud. If you already have a Google account, it requires no setup. In the examples below, I am using Google Colab.

Visual Studio Code

Visual Studio Code is another application that can be used for the same tasks. This is more commonly found since you can use it to do other software engineering tasks.

Popular Sites and Datasets

Yahoo Finance

Yahoo has been around since 1994, so I assume we all know what that is. The company created an API that gives you free access to a wide range of data. You can obtain real-time and historical financial data. This data can be currencies, stocks, bonds, cryptocurrencies, market analysis, and even news!

I will show you a demo of how to use it using a popular Python package called yfinance.

First, we will install the Python package yfinance.

```
!pip3 install yfinance
```

Next, we will import the necessary libraries needed to get data. We need to import yfinance. We will import it as "yf" to make it simpler and faster to write the code.

```
import yfinance as yf
```

In this example, we will focus on the APPLE stocks. We download stock data from January 1, 2020, to December 31, 2022. Next, we will save that data in "apple_data" and use that moving forward.

```
apple_data = yf.download("AAPL",

             start = "2020-01-01",

             end = "2022 - 12 -31")
```

As you can imagine, there is a lot of data in 2 years so by calling ".head()," you can see the first five rows of data, which can be seen in Figure 2-5.

```
apple_data.head()
```

Date	Open	High	Low	Close	Adj Close	Volume
2020-01-02	74.059998	75.150002	73.797501	75.087502	73.249008	135480400
2020-01-03	74.287498	75.144997	74.125000	74.357498	72.536911	146322800
2020-01-06	73.447502	74.989998	73.187500	74.949997	73.114891	118387200
2020-01-07	74.959999	75.224998	74.370003	74.597504	72.771042	108872000
2020-01-08	74.290001	76.110001	74.290001	75.797501	73.941643	132079200

Figure 2-5. *First five rows in data*

Now, we can plot this data to see any trends or patterns. In this example, we will look at the Adjusted Close prices in the "Adj Close" column seen in Figure 2-5.

```
apple_data["Adj Close"].plot(title = "Apple Stock Price
2020 -2022")
```

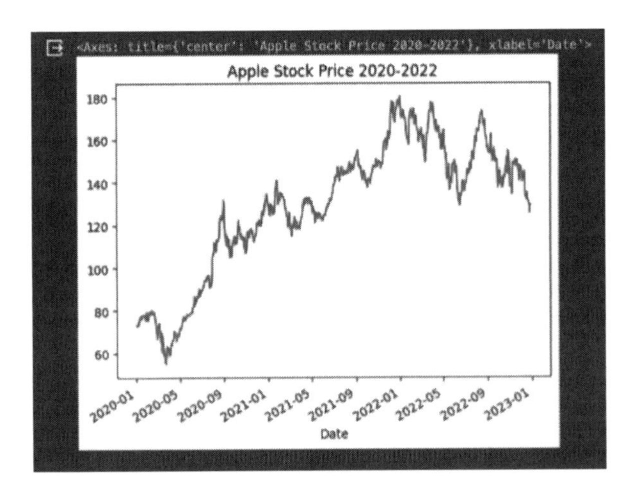

Figure 2-6. *Output of stock prices plot*

Quandl

Quandl is a platform that provides its users with financial and alternative datasets. Alternative datasets are data that include broker notes and company filings. It was founded by Tammer Kamel in 2011 and acquired by Nasdaq in 2017. Over 20 million datasets are available to users that you can get for free or pay for. You can download the data in two ways. One way is in time series, and the other is through a table. Below, I will show you how to retrieve data. This time, we will use an API to get the data. To receive an API key, you will need to create an account at `https://data.nasdaq.com/`.

First, we will install the Python package Nasdaq Data Link.

```
!pip3 install nasdaq-data-link
```

Next, we will import the necessary libraries to get our data. We will only need to import Nasdaq data link.

```
import nasdaqdatalink
```

Finally, we will get all the data for NSE/OIL and output it using our API key. As you can see in Figure 2-7, we have access to oil data from September 30, 2009, to January 4, 2019.

```
nasdaqdatalink.get("NSE/OIL", api_key = "YOUR_API_KEY")
```

	Open	High	Low	Last	Close	Total Trade Quantity	Turnover (Lacs)
Date							
2009-09-30	1096.00	1156.70	1090.00	1135.00	1141.20	19748012.0	223877.07
2009-10-01	1102.00	1173.70	1102.00	1167.00	1166.35	3074254.0	35463.78
2009-10-05	1152.00	1165.90	1136.60	1143.00	1140.55	919832.0	10581.13
2009-10-06	1149.80	1157.20	1132.10	1143.30	1144.90	627957.0	7185.90
2009-10-07	1153.80	1160.70	1140.00	1141.45	1141.60	698216.0	8032.98
...	...	...	...	...	...	...	...
2018-12-31	178.10	179.00	174.35	175.00	174.80	761462.0	1343.75
2019-01-01	175.00	176.40	174.15	175.15	175.75	381570.0	669.18
2019-01-02	175.60	176.20	171.00	172.35	172.40	722532.0	1251.86
2019-01-03	172.80	175.70	171.50	172.00	172.00	698190.0	1212.34
2019-01-04	172.05	174.95	172.05	174.55	174.55	431122.0	743.99

2299 rows × 7 columns

Figure 2-7. *List of available data from NSE/OIL*

Synthetic Data

Synthetic data mimics data in the real world. It is usually created when the actual data is difficult to access because it is expensive or contains sensitive information. This type of data can alleviate business concerns about data privacy by not revealing personally identifiable information. Data leaks of customer information can cause lawsuits and affect brand reputation, hurting a company's bottom line. Generating synthetic data can help create faster turnaround times for product testing and developing prototypes, depending on the situation. Synthetic data generation is the process used to automate or manually create this fake data. This data can be created manually using Excel or a computer algorithm using tools like Python.

A dataset can be fully or partially synthetic. When a dataset is fully synthetic, it has no relation to the real world. However, when the data is partially synthetic, only some of the data is fake. For example, only the name, address, and phone number would be filled with fake data in a spreadsheet regarding home purchases from 2013 to 2023.

There are also situations when there is not enough data. Machine learning models, especially deep learning models, need vast amounts of data to train on. Adding on to a dataset using synthetic data may be necessary when there is a lack of the required data.

Though this type of data is useful, it should only be used when necessary. What you are building will eventually deal with real-world events so you should use real data if you have access to it.

Let's do a quick example of generating synthetic data using Python. In this example, I will create fake income data.

First, we will install the Python package Faker. Faker is a Python library that can generate names, addresses, phone numbers, email addresses, and more!

```
!pip3 install faker
```

Next, we will import the necessary libraries to create our sample data. As mentioned, Faker generates fake data. Pandas is a library that helps us organize and clean data. We will import it as "pd" to make it simpler and faster to write the code. Random is a library that generates random numbers. We need this to generate random incomes.

```
import faker import Faker
import pandas as pd
import random
```

Faker can generate data in various languages. I am specifically making sure the information generated is related to the United States by adding "en_US". This will make phone numbers and names of people in formats known to people in the United States.

```
fake_data = Faker("en_US")
```

Now, we will generate the data. We will create three lists. The people list will contain random names, the phonenumbers list will contain random numbers, and the income list will contain random incomes.

```
people = []
phonenumbers = []
incomes = []
```

In this example, we will create a list of ten people with varying incomes by generating a loop.

```
for i in range(10):
        person_name = fake_data.name()
        phone_number = fake_data.phone_number()
        income = random.randrange(40000, 200000)
        people.append(person_name)
        phonenumbers.append(phone_number)
        incomes.append(income)
```

Finally, we will put the data into a structured format so that we can read it by placing the data into a DataFrame. A DataFrame organizes data into a table with rows and columns like a spreadsheet.

```
sample_data = pd.DataFrame()
sample_data["People"] = people
sample_data["Phone Number"] = phonenumbers
sample_data["Income"] = incomes
```

Below is an output of what the data looks like. As you can see from Figure 2-8, the data is structured like a spreadsheet.

	People	Phone Number	Income
0	Anthony Avila	+1-227-874-2486x878	185205
1	Kimberly Cruz	960.649.0885	149386
2	Cassidy Adams	001-285-410-5956x4515	168926
3	Justin Holmes	(695)708-9488x072	54419
4	Jasmine Lopez	554-924-9009	105524
5	Amber Sanchez	(851)225-3475x80479	184795
6	Gregory Rodriguez	825.709.8450x5884	62139
7	James Greene	595-814-0918	163618
8	John Reed	001-938-957-3118	177171
9	Colleen Martin	8658095302	181263

Figure 2-8. *Output of the synthetic data in a DataFrame*

You can also convert the DataFrame into an actual spreadsheet. Below, I have converted the DataFrame into a CSV using ".to_csv".

```
sample_data.to_csv("synthetic_data.csv", index = False)
```

Telemetry

Telemetry is the collection of raw data in real time from a device or application. Examples of telemetry are monitoring data from space crafts, sensory data from your car to get the speed of your vehicle, and user activity on mobile/web applications.

Since this data is raw, we will have to clean and structure the data in a way that will be useful for us to analyze. We will discuss ways to do that in the next chapter. Once this data is cleaned, it can be viewed on a dashboard to monitor how a device or application is working in production. This data can also be fed back to the system to improve it further with time.

Case Studies and Real-World Examples

Let's delve into examples that showcase how different data collection methods, including the use of real and synthetic data, can be applied in different industries.

Obtaining Real-World Data

Let's explore practical examples of sourcing real-world data to understand how it can be used to do analysis and inform decision-making.

Retail Sales Forecasting

Scenario: A retail chain aims to improve inventory management and sales forecasting.

Data Collection Method: The company collects historical sales data, customer transactions, and external factors such as economic indicators and weather patterns.

Outcome: By analyzing this data, the retailer can predict future sales trends, optimize inventory levels, and better strategically plan marketing campaigns.

Financial Fraud Detection

Scenario: A bank wants to enhance its fraud detection system to protect customers.

Data Collection Method: Transaction data, including user behavior, location, and transaction history, is collected in real time.

Outcome: By analyzing patterns in the data, the bank can identify anomalies, detect potential fraud, and enhance security measures to safeguard customer accounts in the future.

Supply Chain Optimization

Scenario: A manufacturing company wants to optimize its supply chain to make it more efficient and cost-effective.

Data Collection Method: Data is collected from suppliers, production lines, and distribution networks.

Outcome: By analyzing the data, the company can identify any bottlenecks, reduce the time it takes to deliver an order after receiving it, and enhance the overall supply chain performance.

Obtaining Synthetic Data

Let's explore practical examples of reasons to generate synthetic data to complement and enhance real-world datasets.

Preserving Privacy in Healthcare

Scenario: Researchers need to share healthcare data for collaborative studies while ensuring patient privacy.

Reason for Utilizing Synthetic Data: Instead of sharing actual patient records, synthetic data is generated to simulate realistic patient data to protect the privacy of individuals.

Outcome: Researchers can use synthetic data for collaborative analysis, maintaining the confidentiality of patients and complying with privacy regulations.

Image Recognition

Scenario: A company is developing an image recognition model but lacks a sufficient diverse dataset.

Reason for Utilizing Synthetic Data: In order to augment the real-world dataset, they currently have, synthetic images are generated to simulate variations in lighting conditions, angles, and backgrounds.

Outcome: The model trained on a combination of real and synthetic data demonstrates improved performance and generalization to new unseen scenarios.

Ethics and Liabilities

In this chapter, I have shown you various ways to obtain data. However, it is essential to obtain data ethically and responsibly. You should always follow data protection regulations, consider all privacy issues, and get user consent when possible. Below are a few laws to keep in mind.

CCPA: California Consumer Privacy Act

The CCPA is a law that allows any resident in California to see all the information a company has saved on them. Residents can also see a list of third parties with whom their data is being shared. This also allows residents to sue companies if there is a violation of privacy guidelines, even if the company has not been hacked. The company does not need to be in the United States to be affected. Companies that make at least 25 million dollars in revenue, have at least 50,000 users, or collect half of their sales from selling user data are vulnerable to this law.

GDPR: General Data Protection Regulation

GDPR is a law that governs how the personal data of individuals in the European Union (EU) may be processed and transferred. Any data that is collected within or transferred from any of the countries in the EU is affected. If the laws are violated, the fees are very high. You will have to pay whichever is highest, between 20 million euros or 4% of global revenue.

HIPAA: Health Insurance Portability and Accountability Act

HIPAA is a federal law that protects sensitive health information from being disclosed without consent or knowledge. This covers all individually identifiable health information in any form or media.

Enhancing Transparency and Understanding Using Dataset Cards

Dataset cards, also known as Data Cards, are an essential tool for documenting and understanding the characteristics of datasets. They offer a structured format to capture critical information about a dataset, making it easier for people to assess its suitability for various use cases. A dataset card is a concise summary that provides detailed information about a dataset, including its capabilities, limitations, and intended use. It typically includes, but not limited to, key sections such as

- Summary: A brief overview of the dataset, its purpose, and what tasks it supports (e.g., summarization, etc.)
- Languages: Information about the languages represented in the dataset (e.g., English, French, etc.)
- Structure: Overview of how the data is structured, including examples and descriptions of the fields in the data
- Creation: Information on how the dataset was curated and collected
- Source Data: Details on where the original data came from, including collection methods and sources
- Considerations: Ethical and social impacts, biases, known limitations, and potential risks of using the data

Why Are Data Cards Important?

Dataset cards enhance transparency by providing clear documentation that helps people understand the data they are working with. By including dataset cards, you ensure that data is being used appropriately and save people a lot of time when trying to understand how to use the data. Dataset cards are valuable for anyone involved in data analysis, data science, and machine learning. It offers a standardized way to communicate important information. I encourage you all to create dataset cards when sharing data. Sites like Hugging Face and Kaggle usually have this documentation.

Conclusion

To conclude, it is evident that data is the foundational pillar for data analytics, data science, and artificial intelligence. We have explored the diverse types of data and the various acquisition methods, from web scraping, accessing public datasets, tapping into the power of APIs, and creating your own data.

Understanding data, its type, and how to acquire it forms the crucial first step in your journey. The skills and knowledge you have gained in this chapter lay the groundwork for the exciting adventures ahead as you venture further into the world of data analytics, data science, and artificial intelligence.

We have highlighted the importance of ethical considerations, which should remind us that responsible data collection and usage are non-negotiable. It is important to remember that the landscape of data is ever-evolving, so staying curious, ethical, and diligent in your data practices will ensure that you remain at the forefront of this dynamic field. Embrace the complexity, celebrate the diversity of data, and let your passion for discovery guide you through the intricate tapestry of data analytics, data science, and artificial intelligence.

As you enter the next chapter, carry forward the insights and understanding you have gained, ready to apply them in practical, real-world scenarios. The world of data is vast and full of potential - armed with knowledge, curiosity, and a strong ethical compass, you are now ready to navigate confidently.

CHAPTER 3

ETL Pipeline

Ensuring the proficiency of machine learning models is contingent upon the provision of high-quality training data. The accuracy and effectiveness of predictions directly hinge on the caliber of the dataset used during the model's training phase. A well-constructed and maliciously curated dataset serves as the bedrock for machine learning algorithms to discern patterns, grasp intricate relationships, and comprehend nuance information. Consequently, the resultant models are better equipped to make predictions that align with the underlying complexities inherent in the data.

Furthermore, the significance of good quality data extends beyond the initial training phase. As machine learning evolves and encounters new data, the reliability and relevance of ongoing predictions are intrinsically tied to the continuous availability of accurate and representative data. Maintaining a robust data infrastructure ensures that machine learning models remain adaptable and responsive to changes in the environment, yielding predictions that reflect the most current insights and patterns.

In essence, the symbiotic relationship between machine learning models and high-quality data underscores the pivotal role that data quality plays in the efficacy, adaptability, and ongoing success of these intelligent systems.

M. Attobrah, *Essential Data Analytics, Data Science, and AI*,
https://doi.org/10.1007/979-8-8688-1070-1_3

Introduction to ETL

ETL (Extract, Transform, and Load) is the foundation of creating efficient machine learning algorithms. ETL is a three-step data integration process (as seen in Figure 3-1) where data is extracted from one or multiple data sources, transformed (cleaned, scrubbed, formatted), and loaded into a database.

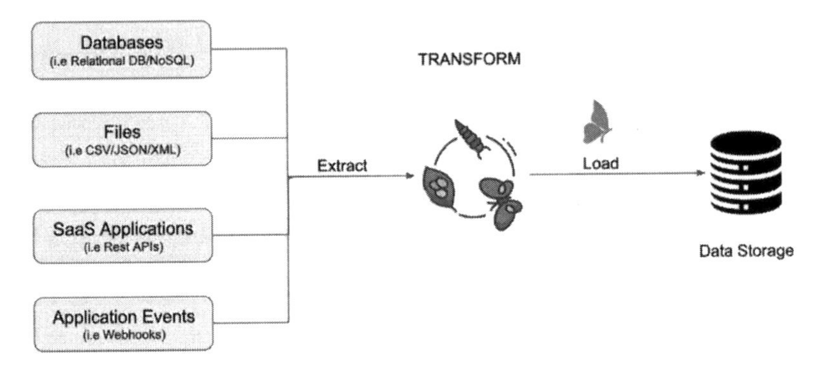

Figure 3-1. *Image of the ETL pipeline*

There are various types of ETL tools available with nice user-friendly interfaces that you can use to complete these tasks. However, Python has many open source ETL tools that can get the job done as well. One very popular tool within the data scientist community is Pandas.

Within the transformation step of ETL, exploratory data analysis (EDA) and data cleaning are important tasks to be performed to identify and remove things such as incorrect, corrupted, or duplicate and inconsistent data.

Importance of ETL

The ETL process plays a pivotal role in ensuring that data, often originating from diverse and disparate sources, is harmonized into a uniform structure. As businesses accumulate information from various channels, systems, and applications, the data may exist in different formats and structures. ETL serves as the bridge that normalizes this heterogenous data, resolving inconsistencies and transforming it into a standardized format. By enforcing a consistent structure, companies can eliminate redundancies, errors, and discrepancies that may arise from disparate data sources. This standardization not only facilitates seamless integration but also enhances the accessibility of information across the organization.

Moreover, ETL empowers businesses to break down data silos by consolidating information from different departments or systems into a centralized data warehouse or database. The consolidation simplifies data retrieval and analysis, providing a comprehensive and unified view of the organization's operations. As a result, teams across various departments rely on a single, reliable source of truth, fostering collaboration and enabling more efficient decision-making processes.

The standardized and centralized data obtained through ETL processes serves as the foundation for data-driven decision-making. With a consolidated dataset in a consistent format, businesses gain the ability to extract meaningful insights, identify trends, and derive actionable intelligence. This newfound clarity enables stakeholders to make informed decisions, backed by accurate and up-to-date information. From strategic planning to operational adjustments, companies can align their actions with a deeper understanding of their data, fostering a culture of evidence-based decision-making.

Furthermore, ETL facilitates historical data tracking, allowing companies to monitor and evaluate their performance over time. By analyzing historical trends and patterns, businesses can identify areas for improvement, assess the impact of strategic initiatives, and refine their operations. The continuous loop of data extraction, transformation, and loading not only supports day-to-day decision-making but also provides a foundation for long-term business optimization, ensuring adaptability in an ever-evolving business landscape. In essence, ETL empowers organizations to not only understand their current state but also to evolve and thrive in a data-driven future. Below is a summarization of some key values ETL brings:

- Singular Perspective

  Obtaining a comprehensive understanding of your business presents a challenge in today's business environment, as data is often dispersed across various systems and applications. Recognizing the significance of ETL in data integration enables you to utilize ETL, facilitating the consolidation and analysis of data from diverse sources to achieve a more holistic view of your business.

- Automated Processes

  ETL streamlines repetitive data processing tasks to enhance analysis efficiency. ETL tools have the capability to automate data migration and integrate regular or on-demand data updates. This enables data engineers to allocate less time to routine tasks such as data movement and transformation.

- Data Governance

  Data governance focuses on the effectiveness, accessibility, consistency, integrity, and security of the data. ETL plays a role in supporting data governance by introducing a layer of abstraction between the source and target systems, thereby preserving data security and lineage. The significance of data governance is becoming more evident as it fosters data democracy, enhancing data accessibility and making it available to all stakeholders for business analysis.

- Scalability

  With the escalation in both data volume and complexity, the significance of ETL in data integration becomes progressively crucial for companies aiming to maintain competitiveness. The ability to scale up ETL pipelines ensures the seamless extraction, transformation, and loading of substantial data volumes. Consequently, the company can effectively handle the growing influx of data without necessitating frequent upgrades, ensuring timely availability for analysis and reporting and maintaining system functionality. Scalability can be attained through various methods, including distributed processing, parallel processing, data partitioning, and the utilization of cloud solutions.

- Rapid Retrieval

  Efficiently retrieve a multitude of integrated and refined data as a business user to support decision-making. As ETL tools handle the bulk of processing during data transformation, the imported data is already in a usable format upon entering the data store. With a well-defined perspective, your business intelligence applications can generate reports more swiftly, as they are spared the need for intricate complications, joining records, or the maintenance of formatting standards when querying the database.

- Minimization of Human Errors

  Despite diligent efforts, manually managing data exposes you to risk of errors. A small oversight in the early stages of data processing can lead to significant consequences. This is because a single mistake can trigger a chain reaction of larger errors, perpetuating a continuous cycle. Recognizing the significance of ETL solutions, many organizations have embraced automation to minimize manual involvement and associated risk of errors, ensuring the delivery of high-quality data.

- Enhanced Quality and Uniformity

  ETL aids in the detection and correction of errors, inconsistencies, and other data issues using data cleaning and transformation techniques. This enhances the quality, reliability, accuracy, and dependability of decision-making.

ETL Tools and Techniques

Over the years, the landscape of ETL has gone a remarkable evolution, driven by the increasing complexity and diversity of data sources as well as the growing demand for efficient data integration solutions. Initially, ETL processes were often custom coded, which required significant development effort and expertise. However, as the need for more scalable, user-friendly, and adaptable solutions emerged, a variety of specialized ETL tools entered the market.

These tools vary widely in their functionalities, catering to different business needs and technical requirements. Enterprise ETL solutions provide comprehensive features and robust support, making them suitable for large-scale data integration projects. These tools often come with advanced capabilities such as data quality management and real-time data integration addressing the intricate challenges posed by modern data environments.

In parallel, the rise of open source ETL tools has democratized the data integration landscape, offering cost-effective alternatives with a strong community-driven development model. Tools like Apache Kafka and Keboola exemplify the flexibility and customization possibilities that open source solutions bring to the table. These tools not only provide powerful ETL capabilities but also encourage collaboration and innovation within the community.

Additionally, the demand for tailored solutions led to the development of custom ETL solutions, where organizations build in-house tools to precisely fit their unique requirements. This approach is particularly common when businesses have specific data integration needs or workflows that are not readily addressed by off-the-shelf solutions. While custom solutions offer a high degree of customization, they also require ongoing maintenance and may lack some of the out-of-the-box features present in commercial ETL tools.

In essence, the diverse array of ETL tools available today reflects the dynamic nature of the data integration landscape. Whether organizations opt for enterprise solutions for comprehensive support, leverage the collaborative power of open source tools, or develop custom solutions for specialized needs, the evolving ecosystem of ETL tools continues to play a crucial role in helping businesses unlock the future potential of their data.

- Enterprise Software ETL

  In the expansive domain of data integration, numerous companies offer commercial ETL software, distinguished by their user-friendly Graphical User Interfaces (GUIs). These GUIs are designed to simplify the complexity of designing and executing ETL pipelines, providing an intuitive drag-and-drop interface that eliminates the need for intricate coding.

This visual approach not only democratizes the ETL process, enabling a broader range of professionals to actively participate, but also enhances the comprehension of the entire workflow. Commercial ETL tools with GUIs often come equipped with pre-built connectors and transformations, accelerating pipeline development and facilitating swift adaptations to evolving data requirements.

With features such as scheduling and monitoring, these tools streamline data integration, contributing to the overall efficiency of organizations in managing diverse and complex data ecosystems. Below are examples of enterprise software ETL solutions:

  - Informatica PowerCenter
  - IBM InfoSphere DataStage
  - Oracle Data Integrator

- Open Source ETL

  Commercial ETL software often comes with a significant price tag, making it a considerable investment for organizations seeking advanced data integration capabilities. However, in contrast, open source tools provide cost-effective alternatives as they are freely available to use. This accessibility democratizes the utilization of ETL capabilities, allowing businesses with budget constraints to benefit from powerful data integration tools. One of the key advantages of open source ETL tools lies in their inherit flexibility. Users with programming expertise can delve into the source code, customizing and tailoring the tool to meet specific requirements. This level of adaptability is particularly advantageous for organizations with unique data processing needs, as it enables them to fine-tune the ETL processes according to the intricacies of their data landscape. The open nature of the source code fosters a collaborative community, encouraging knowledge sharing and innovation as users contribute to the enhancement of the tool's features and functionalities. Overall, the availability and the adaptability of open source ETL tools not only mitigate financial constraints but also empower users to have greater control over the customization and optimization of their data integration workflows. Below are examples of open source ETL solutions:

- Keboola
- Talend Open Studio for Data Integration
- Apache Kafka

- Custom ETL

  For those seeking unparalleled flexibility in their data integration processes, the option of creating custom ETL tools tailored to specific needs becomes an enticing concept. Leveraging programming languages like Python, individuals and organizations can embark on the creation of bespoke solutions that precisely align with their unique requirements. This approach, while offering the utmost adaptability, demands a substantial investment of effort. Building a custom ETL tool entails not only the initial coding but also the creation of comprehensive documentation to ensure clarity and accessibility for future users and developers. The responsibility doesn't end with the initial development; ongoing maintenance is crucial to address evolving data landscapes and changing requirements. Rigorous testing becomes imperative to guarantee the tool's reliability, scalability, and performance. Despite the efforts involved, the creation of custom ETL tools provides unmatched level of control and customization, allowing for seamless integration of data in a manner that aligns precisely with the objectives and intricacies of the organization's data ecosystem. Below are examples of programming languages you can use to create custom ETL solutions:

- SQL
- Spark
- Python

- ETL Cloud Service Tools

  Data within cloud services, such as Amazon Web Services (AWS), Azure, or Google Cloud Platform (GCP), is accompanied by robust and integrated ETL tools native to these platforms. These cloud-based ETL solutions offer seamless and comprehensive approach to handling data workflows within the cloud environment. In AWS, for instance, services like AWS Glue empower users to extract, transform, and load data efficiently, providing not only a scalable and serverless architecture but also automated data discovery and schema evolution. Azure's ETL offerings, including Azure Data Factory, streamline the movement of data across various services, facilitating the creation of data pipelines for diverse scenarios such as analytics, reporting, and machine learning. Similarly, GCP offers tools like Cloud Dataflow, providing a fully managed service for ETL processing with Apache Beam, enabling parallel data processing scale. Leveraging these integrated ETL tools within cloud services simplifies the orchestration of data workflows, optimizes performance, and takes advantage of the scalability elastically inherent in cloud computing environments, ultimately contributing to a more efficient and streamlined data management process. Below are examples of cloud ETL solutions:

- AWS Glue
- Azure Data Factory
- Google Cloud Dataflow

EXAMPLE WITH CODE

The following example I will do using Python. To follow along, you can use software applications like Jupyter Notebooks, Google Colab, or even Visual Studio Code, which are all free.

Pyjanitor is an open source tool in Python used to clean datasets. The advantage of this tool is that it provides an easier-to-read, verb-based method chaining API for common Pandas routines. This helps to make writing code cleaner and easier to remember.

Installation

Install pyjanitor by running the command below:

```
!pip3 install pyjanitor==0.28.1

!pip3 install pandas==2.2.2
```

Implementation

a. Import required libraries.

We will import pandas to extract the data and help with data analysis. Pyjanitor is used to clean data.

```
import pandas as pd
import janitor
```

b. Extract.

In this example, the goal is to provide quality data to a machine learning model of a movie recommendation system so that it can predict relevant recommendations for a user. There are many types of information that can be used. However, to keep this example simple, we will use data that only contains ratings on movies from existing users in our system. This data is in a CSV file called ratings.csv.

We are going to extract the data from the CSV file and put it into a pandas DataFrame so that we can start the transformation process using pyjanitor.

```
rating_data = pd.read_csv("ratings.csv",
skipinitialspace = True, header = None)
```

Our initial data contains 14,051,752 ratings from users on different movies at different times, as seen in Figure 3-2.

```
rating_data
```

	0	1	2	3
0	2022-09-19T19:46:59	147179	voodoo+tiger+1952	4
1	2022-09-19T19:46:59	235248	sazen+tange+and+the+pot+worth+a+million+ryo+1935	4
2	2022-09-19T19:47:01	9465	body+snatchers+1993	2
3	2022-09-19T19:47:01	205897	total+recall+1990	5
4	2022-09-19T19:47:04	41609	portrait+of+jennie+1948	3
...	...	...	...	...
14051777	2022-11-05T01:43:02	598948	klute+1971	5
14051778	2022-11-05T01:43:03	240511	the+godfather+1972	5
14051779	2022-11-05T01:43:03	439162	special+correspondents+2009	3
14051780	2022-11-05T01:43:03	602405	swimming+to+cambodia+1987	4
14051781	2022-11-05T01:43:03	89187	in+this+world+2002	4

14051782 rows × 4 columns

Figure 3-2. *Image of initial movie ratings*

c. Transform.

Now let's do some analysis and cleaning of the data.

First, we will change the column names to make it easier to read by using the pyjanitor rename_column function. Pyjanitor .rename_column() can only change one column at a time so we will be using method chaining to change all four columns.

```
rating_data = (rating_data
                .rename_column(0, 'date/time')
                .rename_column(1, 'user_id)
                .rename_column(2, 'movie_id)
                .rename_column(3, 'rating))
```

We are only expecting ratings to be an integer between 1 and 5. However, through our initial analysis, we see there are 28 unique ratings in our data, as seen in Figure 3-3 in the second row of the last column.

```
rating_data.describe()
```

	date/time	user_id	movie_id	rating
count	14051782	14051782	14051782	14051782
unique	3733123	1428312	27261	28
top	2022-11-01T22:18:41	344352	the+shawshank+redemption+1994	4
freq	22	65	176748	6476617

Figure 3-3. *Image of the summary of columns in data*

Further analysis shows there is a mix of strings and integers that are in the file, as seen in Figure 3-4. For example, 5 was entered as a number and a text. There are clearly some corrupt data within this file that should be removed.

```
rating_data["rating"].unique()
```

```
array([4, 2, 5, 3, 1, '5', '4', '3', '2', '1', 'D', 'E', 'F', 'N4', 'B',
       'v3', 'C', 't3', 'Q4', 'J3', '44', 'W4', '14', '54', '63', 'z5',
       '55', 'd5'], dtype=object)
```

Figure 3-4. *Image of unique ratings*

In Figure 3-5, we see the number of times different ratings appeared. A rating of E was entered 332 times.

```
4     6476617
5     2902433
3     2573853
4      988404
5      443426
3      393134
2      228598
2       35199
1        8195
1        1312
E         332
F         144
D         110
C          11
55          1
W4          1
z5          1
63          1
54          1
14          1
B           1
44          1
J3          1
Q4          1
t3          1
v3          1
N4          1
d5          1
```

Figure 3-5. *Image of the number of times ratings are shown in data*

More analysis shows that the user_id column has a mixture of integers and text, as seen in Figure 3-7. We are expecting user_ids to be only integers. We would also like to have the date and time each rating was entered to be in a datetime format as opposed to text, as seen in Figure 3-6.

```
rating_data["date/time"].apply(type).value_counts()
```

```
<class 'str'>    14051782
```

***Figure 3-6.** Summary type of data date and time is in*

```
rating_data["user_id"].apply(type).value_counts()
```

```
<class 'int'>    12189696
<class 'str'>     1862086
```

***Figure 3-7.** Summary type of data user id is in*

```
rating_data["movie_id"].apply(type).value_counts()
```

```
<class 'str'>    14051782
```

***Figure 3-8.** Summary type of data movie id is in*

```
rating_data["rating"].apply(type).value_counts()
```

```
<class 'int'>    12189696
<class 'str'>     1862086
```

***Figure 3-9.** Summary type of ratings is in*

The user_id and rating columns will need to be converted into integers. We can do this using the pyjanitor .change_type() function. We will also change the format of the date and time from a string to a datetime format. Anything that cannot be converted will be turned into a NaN and will be removed from the file.

All these steps will be completed using method chaining.

```
rating_data = (rating_data
                .change_type("rating", dtype=int, ignore_
                exception="fillna")
                .change_type("user_id", dtype=int, ignore_
                exception="fillna")
                .to_datetime("date/time", format=" ISO8601",
                errors="coerce")
                .dropna()
                .reset_index(drop = True))
```

Let's look at the values within the rating column after completing the step above. Ratings like E, F, G, and Q4 have been removed from our data. Ratings like 1 have increased. Initially we had a number version of 1 that appeared 8195 times and text version that appeared 1312 times. Now there is only one version that appears 9507 times. This indicates we were able to convert text to numbers. Anything that was not a number, like E or Q4, was removed.

Let's look at the first five rows (Figure 3-10) of our data to see how the other columns were reformatted. Rating and user_id were converted to numbers, and date/time was converted to a datetime format.

```
rating_data.head(5)
```

	date/time	user_id	movie_id	rating
0	2022-09-19 19:46:59	147179.0	voodoo+tiger+1952	4.0
1	2022-09-19 19:46:59	235248.0	sazen+tange+and+the+pot+worth+a+million+ryo+1935	4.0
2	2022-09-19 19:47:01	9465.0	body+snatchers+1993	2.0
3	2022-09-19 19:47:01	205897.0	total+recall+1990	5.0
4	2022-09-19 19:47:04	41609.0	portrait+of+jennie+1948	3.0

Figure 3-10. *Image of movie recommendation data*

You may have noticed that user_id and rating are decimals (i.e., floats) instead of whole numbers (i.e., integers). This is because while we were converting text to integers in the previous steps, NaN values do not have an integer version so the data in the entire column are changed in floats. This is similar to how the pandas .to_numeric() function works. To fix this for both columns, we can use change_type and method chaining again.

```
rating_data = (rating_data
                .change_type("rating", dtype=int, ignore_
                exception="fillna")
                .change_type("user_id", dtype=int, ignore_
                exception="fillna")
                .dropna()
                .reset_index(drop = True))
```

Now if we look at the first five rows (Figure 3-11), we can see that data in user_id and rating columns are now in integer format.

```
rating_data.head(5)
```

	date/time	user_id	movie_id	rating
0	2022-09-19 19:46:59	147179	voodoo+tiger+1952	4
1	2022-09-19 19:46:59	235248	sazen+tange+and+the+pot+worth+a+million+ryo+1935	4
2	2022-09-19 19:47:01	9465	body+snatchers+1993	2
3	2022-09-19 19:47:01	205897	total+recall+1990	5
4	2022-09-19 19:47:04	41609	portrait+of+jennie+1948	3

Figure 3-11. *Image of movie recommendation data*

d. Load.

This exports the transformed data into a new .csv file which can be used in a machine learning algorithm.

```
rating_data.to_csv("clean_rating_data.csv")
```

Case Studies and Real-World Examples

Retail Analytics

Scenario: A retail company wants to analyze its sales data to optimize inventory and improve overall business performance.

ETL Process: Extracting sales data from various sources (point-of-sales systems, online platforms), transforming it by cleaning and standardizing formats, and loading it into a centralized data warehouse.

Outcome: The company can now perform advanced analytics, such as predicting demand trends, identifying popular products, and optimizing stock levels.

Healthcare Data Integration

Scenario: A healthcare organization aims to improve patient care by integrating data from Electronic Health Records (EHR), laboratory results, and patient feedback.

ETL Process: Extracting data from different systems, transforming it by ensuring data consistency and privacy compliance, and loading it into a unified database.

Outcome: Enables healthcare professionals to access comprehensive patient histories, improving diagnostic accuracy and personalized treatment plans.

Social Media

Scenario: A social media analytics company wants to analyze sentiment trends to help businesses understand customer opinions about their products or services.

ETL Process: Extracting data from social media platforms, transforming it by applying natural language processing (NLP) techniques to analyze sentiment, and loading the results into a data store for reporting.

Outcome: Businesses can gain insights into customer satisfaction, identifying areas for improvement, and tailor their marketing strategies accordingly.

Fraud Detection in Financial Transactions

Scenario: A financial institution wants to detect fraudulent activities in real time to secure transactions and protect its customers.

ETL Process: Extracting transaction data from various channels, transforming it by applying anomaly detection algorithms, and loading flagged transactions into a fraud monitoring system.

Outcome: Helps in preventing fraudulent transactions, minimizing financial losses, and maintaining the trust of customers.

Smart Home Automation

Scenario: A company in the IoT (Internet of Things) space aims to create a smart home platform that learns user behavior and adapts home automation settings accordingly.

ETL Process: Extracting data from various sensors and devices, transforming it by identifying usage patterns, and loading the insights into a central system to automate home devices.

Outcome: Users experience a personalized and automated home environment based on their habits, optimizing energy usage and enhancing convenience.

Conclusion

In conclusion, the process of obtaining and managing data is a critical foundation for success in the realms of data analytics, data science, and artificial intelligence. From exploring various sources to harnessing the power of ETL tools, the chapter has underscored the importance of robust strategies for obtaining both real-world and synthetic data. Whether delving into commercial ETL software with user-friendly interfaces, leveraging open source tools for cost-effective flexibility, or venturing into the realm of custom ETL solutions, practitioners have a spectrum of options to suit their specific needs. Additionally, the integration of cloud services with built-in ETL tools further amplifies the efficiency of data workflows. As we navigate the ever-evolving landscape of data, the chapter has illuminated the diverse avenues and tools available, empowering readers to navigate the data terrain with confidence, creativity, and a strategic approach to meet the unique challenges and opportunities that data-driven endeavors present.

CHAPTER 4

Exploratory Data Analysis

In the previous chapter, we discussed what an ETL pipeline is, its importance, and how to implement it. What do we do next once we load this data in its desired location? Sometimes, you can have so much data that it may be difficult to understand where to start or finish. It may be difficult to see at first how this data can help your organization. Imagine inheriting a treasure chest from the past with numerous journals. Each journal has stories, cryptic symbols, and maps. As you look through this treasure, you realize the key to unraveling the secrets held within this chest is understanding and connecting the hidden patterns you find. Similarly, exploratory data analysis is like going through a treasure trove of data – each dataset is a collection of numbers, variables, and observations. Just as you would carefully examine the journals to decipher their contents to piece together the narrative of their originating authors, exploratory data analysis involves pulling back the layers to obtain insights, trends, and anomalies lurking hidden beneath the surface.

Introduction to Exploratory Data Analysis

In the vast landscape of data analytics, data science, and artificial intelligence, a crucial step serves as both a compass and a magnifying glass – exploratory data analysis (EDA). Much like a seasoned explorer

M. Attobrah, *Essential Data Analytics, Data Science, and AI*,
https://doi.org/10.1007/979-8-8688-1070-1_4

surveying unchartered territory, EDA delves into datasets and uncovers hidden gems of insight and understanding. In this chapter, we embark on a journey through the realms of data exploration, understanding its significance and transformative power.

The essence of exploratory data analysis lies in its capacity to unveil the stories concealed within the data folds. It is the art of looking closely beyond the surface, seeking patterns, anomalies, and trends that lurk beneath. Creating a toolbox with a diverse arsenal of statistical techniques, visualizations, and domain knowledge allows you to breathe life into datasets that look like it lacks vigor on the surface. You transform rows and columns into narratives of discovery and enlightenment. From the quaint quirks of consumer behavior to the intricate dynamics of financial markets, EDA serves as the conduit through which data speaks its truth.

Moreover, exploratory data analysis is not merely a technical preamble to more sophisticated analysis but an essential voyage of validation and discovery. It is the crucible in which hypotheses are forged, assumptions are challenged, and insights are distilled. Through critical observation of the quality of the data and relationships between variables, EDA provides the evidence-based grounding upon which robust tools for data analytics, data science, and artificial intelligence are constructed. In essence, it is the lighthouse that guides you through the tumultuous seas of uncertainty, illuminating the path toward knowledge and understanding.

Data distributions describe how data points are spread across a dataset. This can significantly influence the methods used during EDA. Common distributions include but are not limited to

- Normal Distribution

  A normal distribution is also known as a Gaussian distribution and is symmetrical, with most data clustered around the mean. It forms a bell-shaped curve, where values away from the mean become increasingly rare to see in the data. In cases like this, standard techniques like getting the mean or standard deviation to summarize data are effective.

- Skewed Distribution

  When data is not symmetrically distributed, it is either right-skewed or left-skewed. This skewness affects the choice of how to summarize the data. For example, the mean can be misleading in skewed distributions. So getting the median should be used to represent the central value more accurately. Visualization methods like histograms or box plots can highlight these patterns.

- Uniform Distribution

  In a uniform distribution, every outcome has an equal chance of occurring, leading to a rectangular shape when visualized. Traditional techniques may not be relevant with this type of data because no single area of the data dominates.

By identifying the data distribution, you can tailor your methods. Techniques that work well for normally distributed data might not be suitable for skewed or uniform data.

Importance of Exploratory Data Analysis

Exploratory data analysis is important for several reasons. Below, I will list a few reasons to consider:

1. Understanding the Data

   It helps scientists and analysts comprehend the structure, content, and relationships within one or more datasets. EDA provides a foundation for further analysis by revealing patterns, trends, and potential biases. It guides the direction of analysis and questions, driving actionable insights.

   One key element of EDA is understanding correlation coefficients, which measure the strength and direction between two variables. For example, Pearson's correlation coefficient measures how closely two variables follow a linear relationship. When two variables have a linear relationship, it means that as one variable increases or decreases, the other does so at a constant rate, forming a straight line when plotted on a graph. A Pearson correlation coefficient ranges from -1 to 1, where values closer to 1 indicate a strong positive linear relationship and values closer to -1 indicate a strong negative linear relationship. A value of 0 suggests no relationship. A positive relationship would be when one variable increases or decreases, the other variable moves in the same direction. A negative relationship is when one variable increases or decreases, the other variable moves in the opposite direction.

   Not all relationships between variables are linear. In some cases, variables may still move in a consistent direction relative to one another, but not at a constant rate. This is called a monotonic relationship. In a monotonic relationship, as one variable increases, the other can either increase or decrease, but not necessarily at a constant rate. For example, the relationship might curve. When dealing with monotonic relationships, Pearson's correlation may not capture the strength of association accurately, which is where Spearman's rank correlation can be used. Spearman's measures the consistency of the relationship rather than its strict linearity.

EDA allows analysts to validate assumptions underlying statistical models and methods. By examining data distributions, correlations, and other properties, scientists can ensure that their analyses are appropriate and trustworthy.

2. Data Quality Assessment

   Assessing the quality of the data grants us the opportunity to find any additional outliers, incorrect data entries, missing values, and inconsistencies we may have missed during the Transformation portion of the ETL pipeline. This ensures the data being dealt with is reliable and accurate for analysis and interpretation.

3. Communication and Data Visualization

   This allows you to communicate the data clearly and concisely to leadership and stakeholders. Visual demonstrations of complex data make it more accessible and understandable to the general public and leadership.

4. Feature Selection and Engineering

   After you have a full understanding of your data, it is easier to understand what data will be necessary for the next step. When it comes to model development, not all models need all your data. The leaner your data is, the more your model can focus on what is important to analyze for whatever you have given it to solve. Even as humans, it is easier to solve a problem if you are only provided information related to your problem instead of giving excess information.

   EDA aids in identifying relevant features or variables for predictive modeling and decision-making tasks. It also inspires the creation of new features through transformations or combinations of the data, which enhances the models' predictive power.

There are three main types of EDA:

1. Univariate Analysis

   Univariate analysis implies what the name suggests. When conducting this analysis, you explore one variable at a time. When dealing with data, it can be overwhelming to understand what is going on, so one approach can be to look at each variable separately. Univariate analysis allows you to summarize and identify patterns in a large quantity of data. You can describe data found in this variable using graphs and charts (e.g., bar charts or frequency distribution tables).

Let's say you are tasked with reviewing your company's inventory database and need to find a better way to modernize its data management process. In order to do this, you need to understand what is already in the data. This first dataset you find looks like the table below.

Product ID	Product Name	Category	Quantity in Stock	Price
0001	Product 1	Electronics	200	124.56
0002	Product 2	Clothing	15	45.21
0003	Product 3	Electronics	35	124.56
0004	Product 4	Sports, Fitness, and Outdoors	100	54.21
0005	Product 5	Entertainment	12	34.56
0006	Product 6	Toys	100	124.56
0007	Product 7	Health	200	4.99
0008	Product 8	School & Office	300	45.21
0009	Product 9	Pets	30	32.01
0010	Product 10	Electronics	100	124.56

At first glance, it may be difficult to identify key insights from this data to understand its usefulness to the company. We can use univariate analysis to dissect this data to get a better understanding. You can focus on any of these columns, but for this example, let's focus on the variable Category.

Category
Electronics
Clothing
Electronics
Sports, Fitness, and Outdoors
Entertainment
Toys
Health
School & Office
Pets
Electronics

We can take this variable to see the frequency distribution of the categories.

Category	Frequency
Electronics	4
Clothing	2
Sports, Fitness, and Outdoors	1
Entertainment	1
Toys	1
Health	1
School & Office	1
Pets	1

Here, we can see that there are more electronic products in our inventory. Next, we can add this information to a bar chart to visualize the data more easily.

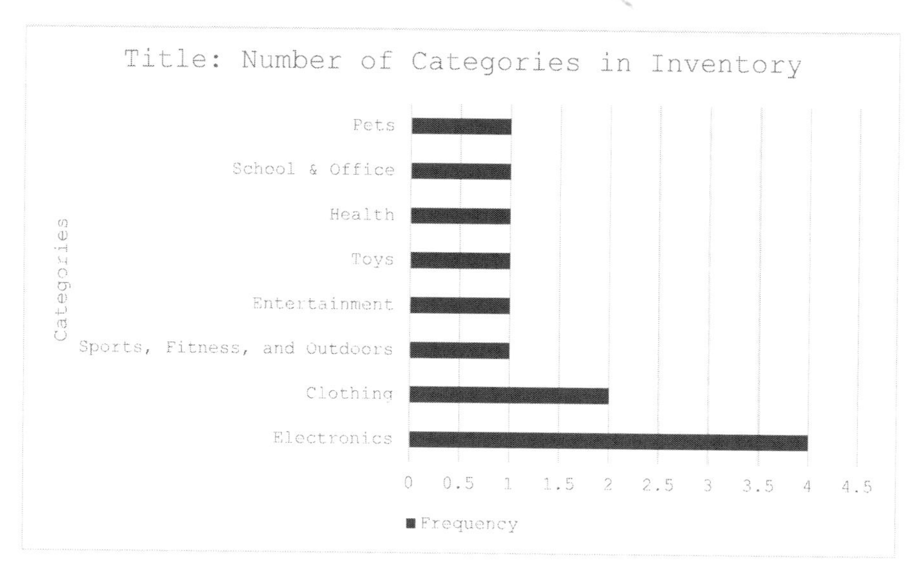

***Figure 4-1.** Bar chart showing the number of categories in the dataset*

Then, we can continue the same individual analysis for each variable within the table above to get an understanding of the data.

2. Bivariate Analysis

Bivariate analysis is the analysis of two variables to determine the relationship between them. Using the example in the table above, we can analyze how the category and quantity left in stock are related. We can also analyze how the product and price are related. For other examples, we can see the relationship between having a pet and emotional well-being. We can also analyze the relationship between money and happiness.

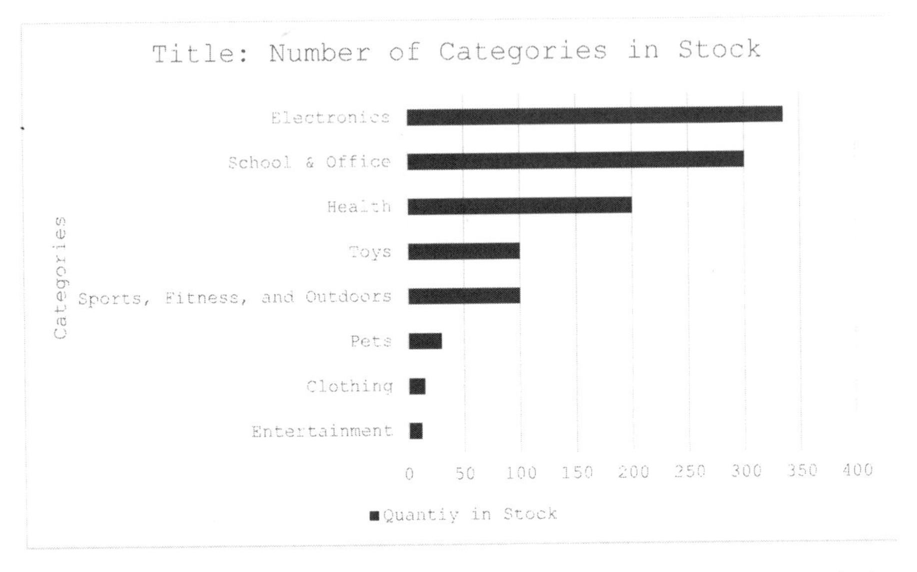

***Figure 4-2.** Bar chart showing the number of categories currently in stock within our inventory*

3. Multivariate Analysis

   Multivariate analysis involves analyzing more than two variables to determine their relationship. It identifies patterns between multiple variables and offers the benefit of considering multiple factors. It examines different aspects of the data that affect the main outcome.

Steps in EDA

The first three steps relate to the ETL process.

1. Gathering Data

   In today's world, vast amounts of data are produced across numerous sectors such as healthcare, sports, manufacturing, and more. Recognizing the value of leveraging data through thorough analysis is a common understanding among businesses. Yet, the effectiveness of this analysis hinges on gathering necessary data from diverse sources like surveys, social media, and data specific to your industry, like medical records. Without acquiring ample and pertinent data, subsequent endeavors cannot commence.

2. Identifying and Comprehending All Variables

   As the analysis journey commences, the spotlight illuminates the trove of available data, brimming with insights waiting to be unearthed. Within the data landscape lie dynamic values encapsulating myriad features or patterns, serving as the bedrock for understanding and enlightenment. The crucial first step involves identifying the key variables that wield influence over outcomes and discerning their potential impact. This pivotal step sets the stage for realizing the ultimate goals of any analysis endeavor.

3. Dataset Cleansing

   This step entails refining the dataset, a task that involves eliminating null values and irrelevant data. This refinement ensures that the data retains only the information essential to the desired analysis. Not only does it save time, but it also reduces the computational resources needed when creating a budget for your project. Preprocessing tackles an array of issues, from flagging null values to identifying outliers and anomalies, ensuring the dataset is primed for analysis.

4. Detecting Correlated Variables

   Discovering correlations between variables provides insights into the relationship between them. A correlation matrix shows the correlation coefficients between variables in your dataset. The values range from -1 to 1, where -1 is considered a weak relationship between variables, 0 is a neutral relationship, and 1 is considered a strong relationship. The correlation matrix offers a comprehensive view of these connections, aiding in the comprehension of crucial interrelationships among variables. This is not always needed in the EDA process but can be useful when selecting what variables to pick when training a machine learning algorithm.

5. Selecting Appropriate Statistical Techniques

   Depending on the nature of the data (whether categorical or numerical), its size, the type of variables, and the objective of the analysis, various statistical tools are utilized. While statistical formulas applied to numerical data are valuable, graphical representations are often more engaging and easier to interpret.

6. Visualization and Examination of Outcomes

   After completing your analysis, the findings must be reviewed thoroughly to ensure accurate interpretations. Observing data trends and correlations between variables provides valuable insights for making necessary adjustments to data parameters and business decisions. A skilled data analyst or data scientist, proficient in all relevant analysis techniques, is essential for this process.

Exploratory Data Analysis Tools and Techniques

There are several tools commonly used to implement EDA. These tools make detecting patterns, spotting anomalies, testing hypotheses, and checking assumptions easier.

Python

In the previous chapter, we provided some examples using Python. Python is currently one of the most popular programming languages for completing artificial intelligence and data science tasks. This is the case because of its easy-to-read syntax, extensive libraries, strong compatibility with cross-platforms, and strong community support. This makes it easier for beginners and experienced programmers to write and understand code and encourages faster development and collaboration.

R

R is another popular programming language. It is primarily used for statistical analysis and data visualizations. It provides a wide range of tools for data manipulation, visualization, and statistical modeling. It is an essential tool for statisticians and data scientists.

Tableau

Tableau is a popular tool for reporting and analyzing large volumes of data. It enables users to create a wide range of charts, graphs, maps, dashboards, and stories for comprehensive data visualizations and analysis. Tableau's robust data discovery and exploration features allow users to quickly answer important questions and make informed business decisions. It requires no prior programming knowledge, making it easier for anyone regardless of technical experience to get started using it immediately.

Power BI

Power BI is another popular tool that requires no prior programming language. Similar to Tableau, it is used for reporting and analyzing data. It provides a suite of business analytics to create data visualizations and share insights across organizations.

EXAMPLE WITH CODE

I will use Python for the following example. You can follow along using free software applications like Jupyter Notebooks, Google Colab, or even Visual Studio Code.

Installation

```
pip install pandas
```

a. Import required libraries.

```
import pandas as pd
import seaborn as sns
```

b. View the data.

The data we are using for this example is on USA housing prices. There are over 4000 rows in this data. The dataset can be found on Kaggle: `https://www.kaggle.com/datasets/fratzcan/usa-house-prices`.

Let's load the dataset using pandas by running the code below:

```
housing_data = pd.read_csv("USA Housing Dataset.csv")
```

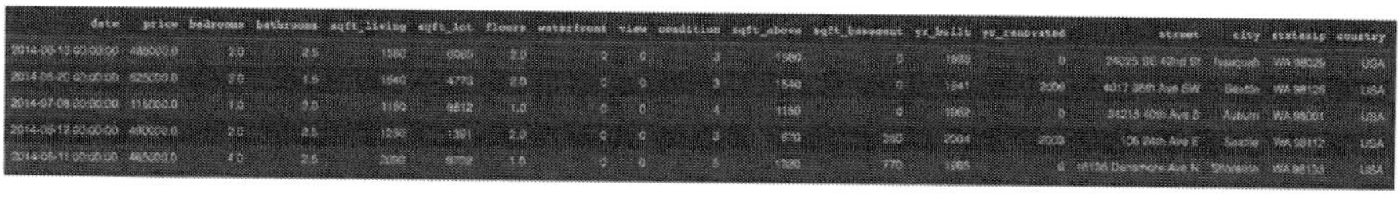

date	price	bedrooms	bathrooms	sqft_living	sqft_lot	floors	waterfront	view	condition	sqft_above	sqft_basement	yr_built	yr_renovated	street	city	statezip	country
[illegible]	[illegible]	3.0	2.5	[illegible]	[illegible]	2.0	0	0	3	[illegible]	0	[illegible]	0	[illegible]	[illegible]	[illegible]	USA
[illegible]	[illegible]	2.0	1.5	[illegible]	[illegible]	2.0	0	0	3	1540	0	[illegible]	[illegible]	[illegible]	Seattle	[illegible]	USA
[illegible]	[illegible]	1.0	2.0	[illegible]	[illegible]	1.0	0	0	4	1150	0	[illegible]	0	[illegible]	Auburn	[illegible]	USA
[illegible]	[illegible]	2.0	2.5	[illegible]	[illegible]	2.0	0	0	3	[illegible]	[illegible]	2004	[illegible]	[illegible]	Seattle	[illegible]	USA
[illegible]	[illegible]	4.0	2.5	[illegible]	[illegible]	1.5	0	0	5	[illegible]	770	[illegible]	0	[illegible]	[illegible]	[illegible]	USA

Figure 4-3. *Image of random samples in the dataset*

The data ranges from May 2014 to July 2014. There are 18 columns within this dataset:

1. date: The date the property was sold
2. price: The selling price of the property
3. bedrooms: Number of bedrooms in the house
4. bathrooms: Number of bathrooms in the house
5. sqft_living: The square footage of the living area
6. sqft_lot: The square footage of the lot

7. floors: The number of floors in the property
8. waterfront: A binary indicator stating whether the property has a waterfront view
9. view: A 0–4 rating of the quality of the views the property has
10. condition: A 1–5 rating of the condition of the property
11. sqft_above: The square footage of the property above the basement
12. sqft_basement: The basement square footage
13. yr_built: The year when the property was built
14. yr_renovated: The year of the most recent renovation of the property
15. street: The property's street address
16. city: The city where the property is located
17. statezip: The state and zip code of the property
18. country: The country of where the property is located

c. Explore the data.

There are many charts, tables, and graphs we can create to help us visualize the data better to understand it. Let's explore the data by viewing what cities have houses over 1.5 million dollars.

```
housing_data[housing_data["price"]>1500000][["city"]].
value_counts()
```

Here we can see there are 17 cities within our dataset that have homes sold for over 1.5 million dollars. Out of this list, Seattle has the most houses, which is 37.

city	count
Seattle	37
Bellevue	21
Mercer Island	17
Medina	8
Kirkland	6
Fall City	2
Issaquah	2
Clyde Hill	2
Shoreline	2
Yarrow Point	2
Kent	1
Carnation	1
Covington	1
Redmond	1
Sammamish	1
Tukwila	1
Woodinville	1

Figure 4-4. *Image of lists of cities with the most expensive houses in the country and the number of houses in that city with that price*

d. View the correlation between variables.

   There are 18 columns in this table, but to keep this example simple, we will only focus on six:

   - price
   - bedrooms
   - yr_built
   - floors
   - waterfront
   - sqft_living

You can use other variables, but you may need to clean the data and reformat it to be able to use them.

```
correlation_matrix = housing_data[["price,"bedrooms",
"yr_built","floors","waterfront","sqft_living"]].corr()
sns.heatmap(correlation_matrix, annot=True)
plt.show()
```

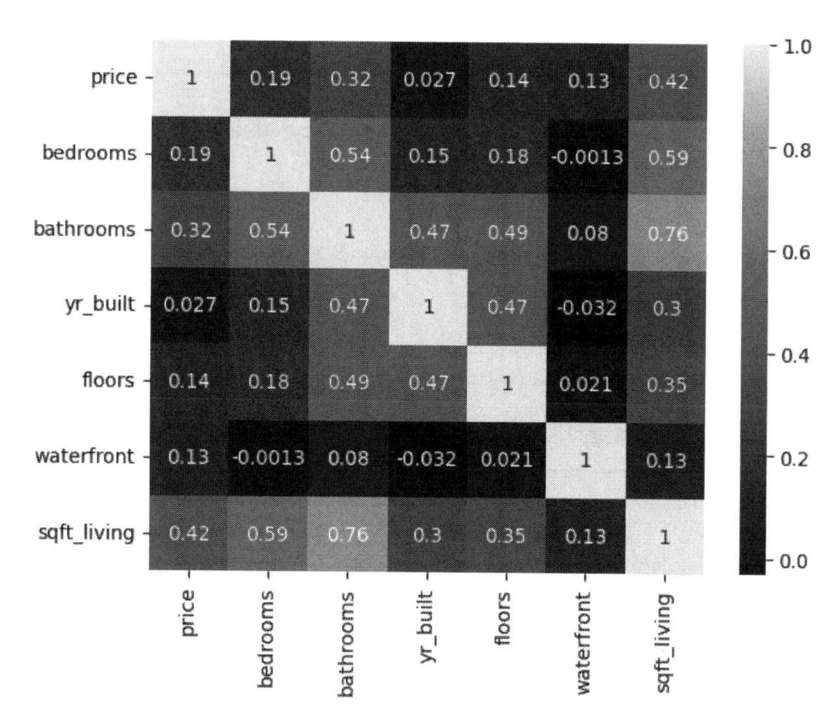

***Figure 4-5.** Correlation matrix showing the relationship between a few variables within the housing dataset*

In this correlation matrix, variables closer to 1 have a strong relationship, and variables closer to 0 have a weak relationship. Here we can see on the first row that price is most affected by sqft_living compared to the other variables. It is least impacted by yr_built. This may be a surprising find to some since one could assume that the year a house was built would have a higher influence on a person's willingness to buy the property. The relationship between price and sqft_living is 0.42, while the relationship between price and yr_built is 0.027.

Choosing the Right Visualizations for EDA

Choosing the right visualization technique is crucial for effectively interpreting data during EDA. Different visualizations highlight various aspects of data, making it easier for you to draw insights and detect patterns.

- Box Plots

  Box plots are helpful in understanding the distribution of a dataset. They display the median, quartiles, and potential outliers. This is helpful when comparing distributions across multiple groups or categories.

- Histograms

  These are ideal for visualizing the frequency distributions of a single variable, allowing analysts to quickly see the shape of the data (e.g., whether it's normally distributed or skewed).

- Heatmaps

  Heatmaps are effective for visualizing relationships between multiple variables, particularly through a correlation matrix, as seen in Figure 4-5. This type of visualization uses color gradients to indicate the strength of correlation between pairs of variables, making it easy to detect strong or weak relationships at a glance.

- Scatter Plots

  Scatter plots are typically used to examine relationships between two numeric variables. They are typically helpful in spotting trends, patterns, and possible linear or nonlinear relationships.

- Bar Charts

  Bar charts are a simple yet powerful tool for comparing different categories. They work well for both categorical and discrete data, allowing analysts to compare quantities across groups or time periods. For example, bar charts can show sales data across various product categories or regions.

- Pie Charts

  Pie charts are useful when you want to represent the proportions of a whole. Each slice of the pie represents a category's contribution to the total. However, pie charts work best when there are a limited number of categories, and the differences between them are significant; otherwise, it will be hard to interpret.

- Line Charts

  Line charts are commonly used for tracking changes and trends over time. They are particularly effective when working with time series data, such as stock prices, weather patterns, or sales performance across months. Line charts make it easy to visualize continuous data and detect upward and downward trends over time.

Each visualization serves a different purpose, and selecting the right one depends on the nature of the data and the questions being asked. For example, box plots are ideal for comparing distributions, and heatmaps are better suited for assessing correlations across multiple variables. Bar charts and pie charts are great for comparing categorical data, and line charts are great for time-based trends. By choosing the appropriate visualization, you can more effectively explore the data and uncover actionable insights.

Use Cases of EDA for Business

Practical examples of how EDA can be implemented provide invaluable insights into how EDA can be applied in practice across various domains. Below are a few examples to illustrate the importance and effectiveness of EDA.

Retail Industry

In the retail sector, EDA can be used to analyze customer purchase behavior, optimize product placement, and identify trends. For instance, a retail company like Macy's may perform EDA on sales data to uncover seasonal patterns, identify bestselling products, and segment customers based on purchasing habits. This information can inform inventory management strategies and marketing campaigns.

Step 1: Collect Data

A retail company like Macy's collects data on sales transactions, including customer demographics, product purchases, dates of transactions, and total sales amounts.

Step 2: Data Cleaning

You can clean the data by handling missing values, removing duplicate entries, and ensuring consistency like product names and customer demographics.

Step 3: Exploration

You can create descriptive statistics and visualizations such as bar charts to understand the sales distribution across product categories and histograms to explore the frequency of purchases across seasons.

Using time series analysis and line charts, you can identify seasonal patterns in sales (e.g., peak sales during holiday seasons). You can also use box plots to analyze variations in sales data by region or customer age group.

Clustering techniques can be used to group customers based on purchasing behavior. Scatter plots and heatmaps can be used to visualize relationships between demographics (e.g., age, income) and purchasing frequency.

Step 4: Actionable Insights

Based on the insights, Macy's can optimize inventory for high-demand products during peak seasons and create targeted marketing campaigns for different customer segments.

Healthcare

EDA can be key in analyzing patient data, identifying risk factors, and improving patient outcomes. For example, a hospital can conduct EDA on electronic health records to identify correlations between patient demographics, medical history, and treatment outcomes. This analysis can help healthcare providers tailor treatments and interventions to individual patient needs. They can also use EDA to analyze the number of patients admitted to the hospital annually. This analysis can help hospitals understand the staffing requirements needed throughout the year.

Step 1: Collect Data

A hospital collects patient data, including demographics (e.g., age, gender), medical history, diagnosis, and treatment outcomes.

Step 2: Data Cleaning

The hospital ensures the data is accurate by removing duplicates and filling in missing values where appropriate.

Step 3: Exploration

Using correlation matrices and heatmaps, analysts examine how various patient demographics correlate with treatment outcomes, identifying which factors (e.g., age, preexisting conditions) are strongly associated with recovery times.

Scatter plots and box plots can be used to explore how different risk factors (e.g., high blood pressure, smoking) are distributed among patients with different outcomes, revealing which groups are at higher risk.

The hospital can also analyze yearly admission rates, using line charts and histograms to forecast staffing needs and resource allocation throughout the year.

Step 4: Actionable Insights

The hospital can use some of this analysis to tailor treatment plans to patient risk profiles. For example, patients with certain risk factors may receive more extensive monitoring or preventative care.

Finance

EDA can be utilized in the finance industry for risk management, fraud detection, and investment analysis. Investment firms can perform EDA on financial market data to identify trends, correlations, and anomalies that could impact investment decisions. Similarly, banks can use EDA techniques to detect suspicious transactions and prevent fraudulent activities.

Step 1: Collect Data

A bank collects transaction data, including transaction amounts, account details, locations, and timestamps.

Step 2: Data Cleaning

The data is cleaned by removing any inconsistencies.

Step 3: Exploration

Using box plots and scatter plots, you can look for outliers in transaction amounts or frequency. Unusual spikes in activity or geographically distant transactions may signal fraud.

Heatmaps are used to analyze correlations between transaction characteristics, such as transaction frequency and location, to detect suspicious patterns that deviate from the norm.

Step 4: Actionable Insights

The bank can flag anomalies for further investigation and use machine learning models trained on the insights from EDA to assist in automating fraud detection in real time.

Manufacturing

In manufacturing, EDA can be used to optimize production processes, improve quality control, and reduce downtime. For example, a manufacturing company may analyze sensor data from production equipment to identify factors contributing to machine failures or defects. This analysis can help prioritize maintenance tasks, minimize production disruptions, and enhance overall efficiency.

Step 1: Collect Data

A manufacturing company collects sensor data from machines, including temperature, vibration levels, and speed.

Step 2: Data Cleaning

You clean the data, ensuring that timestamps and sensor data are aligned across machines.

Step 3: Exploration

Using line charts, you can track performance over time to identify when and why a machine's performance deviates from the norm (e.g., gradual increase in temperature before failing).

Scatter plots and heatmaps are used to find correlations between machine failure events and specific sensor readings. This analysis might reveal that machines that exceed a certain temperature threshold are more likely to fail.

Step 4: Actionable Insights

Based on these findings, the company can implement predictive maintenance schedules, repairing or replacing parts before machines fail, minimizing downtime, and improving production efficiency.

Conclusion

In conclusion, EDA is an important step in data analytics, data science, and artificial intelligence. It helps uncover patterns, trends, and relationships within a dataset. EDA involves various techniques, including data visualizations and statistical analysis, which together enable you to gain a deeper understanding of the data. Tools such as Python, R, and Power BI provide the necessary functionality to effectively perform EDA. This sets the stage for sophisticated visualizations and predictive analysis.

By utilizing EDA, we can identify anomalies and understand correlations between variables. The insights gained during this phase are invaluable for making informed decisions, refining hypotheses, and ensuring the accuracy and relevance of subsequent analysis.

Through EDA, data reveals a story. This guides analysts, scientists, engineers, and stakeholders toward meaningful and actionable insights that drive better business decisions, scientific discoveries, and innovative solutions across various fields. As we move forward, embracing the principles and practices of EDA will continue to be essential in unlocking the full potential of data. Next, we will talk about machine learning models.

CHAPTER 5

Machine Learning Models

Artificial intelligence (AI) is about making computer systems smart enough to do tasks that usually require human thinking, like understanding speech, making decisions, and spotting patterns. AI is an umbrella term that includes different technologies, such as machine learning and natural language processing.

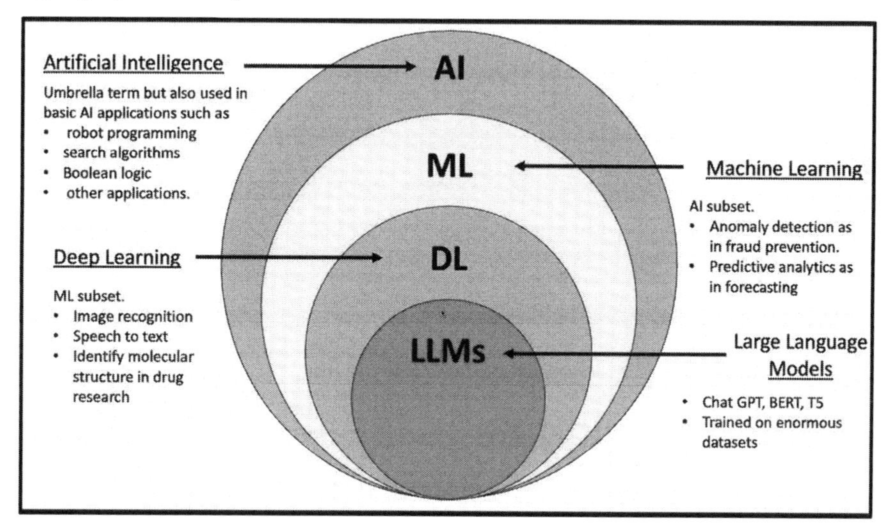

Figure 5-1. *Image from the "AI-ML-DML-GEN AI - LLM sorting according to prices" blog in LinkedIn Pulse, showing the hierarchical relationship between artificial intelligence (AI), machine learning (ML), deep learning (DL), and large language models (LLMs) ([Untitled photograph of relationship between AI, ML, and DL], 2024, April 23)*

M. Attobrah, *Essential Data Analytics, Data Science, and AI*,
https://doi.org/10.1007/979-8-8688-1070-1_5

In this chapter, we will focus on machine learning. Machine learning models represent a pivotal component in the realm of artificial intelligence. These models empower systems to learn from data, identify patterns, and make decisions with minimal human intervention. They have transformed industries by providing unprecedented capabilities in automation and prediction. The ability to automate complex processes and derive actionable insights from vast datasets has revolutionized numerous sectors, from healthcare and finance to retail and transportation.

Machine learning is connected with the fields of data analytics and data science, which focus on collecting, processing, and analyzing data, as discussed in the previous chapters. Data science provides the tools and techniques to extract meaningful insights from raw data, while machine learning leverages these insights to make predictions and drive decision-making. The role of data analytics becomes particularly evident when analyzing large datasets, identifying trends, and ensuring that we have representative data that the model will encounter in the real world during the training phase. This helps in creating models that not only predict but also understand the intricate relationships within the data.

In this sense, data science, data analytics, and machine learning go hand in hand: data analytics and data science help us understand and interpret the data, while machine learning helps us act on it by making predictions and automating decisions.

Ensuring the proficiency of machine learning models is contingent upon the quality of the training data. A well-constructed and curated dataset serves as the bedrock for machine learning algorithms to discern patterns, grasp intricate relationships, and comprehend nuance information. Consequently, the resultant models are better equipped to make predictions that align with the underlying complexities inherent in the data.

Introduction to Machine Learning Models

Machine learning models are the cornerstone of modern data science. They drive advancements across a wide range of industries by enabling components to learn from data and make decisions or predictions without being explicitly programmed. These models can automatically improve their performance with experience, making them compelling tools for complex problems.

Before we continue, let me explain the difference between machine learning models and algorithms. Machine learning algorithms and models are incorrectly used interchangeably, which can be confusing for beginners.

An algorithm is a set of steps designed to accomplish a task. In machine learning, algorithms are procedures that run on datasets to recognize patterns and rules. Think of the algorithm as the process and the model as the final product that results from the process. The model learns from the dataset using the algorithm. Another way of saying it is the model is fitted to the dataset. The output of all of this is a machine learning model that captures these patterns. You can think of a model like an application or program that you can use to input data to make predictions or decisions. The algorithm is what enables the model to learn, but once the learning is done, the model stands alone as the tool for making future predictions or decisions on new, unseen data.

These models create predictions that can help make data-driven decisions for your business.

Machine Learning Model = Data + Algorithm

At its core, machine learning involves training models on a dataset using algorithms to recognize patterns and relationships within the data. This process enables the model to make predictions or decisions when presented with new, unseen data. Depending on how well the model was trained, the model will make predictions with a certain level of accuracy and confidence. There are several types of machine learning algorithms, each suited to different kinds of tasks:

1. Supervised Learning

   Machine learning models are trained using labeled data. This means the data is tagged with the correct answers. Think of this like the model is learning with a teacher. The data contains the inputs and the correct outputs to help the model learn faster.

   Let's say you want to train a machine learning model to predict the value of a used Lamborghini.

   You would have a dataset used to train the model with known input and output values. For example, the data could include

   - Mileage
   - Model (e.g., Sedan, SUV, etc.)
   - Color
   - Fuel Type (e.g., Diesel, Hybrid, Electric, Gasoline)
   - Pricing

   The inputs to the model would be mileage, model, color, and fuel type, and the output would be pricing.

   Once the model has been fitted to the dataset or, in other words, has learned from this dataset, a trained machine learning model using the supervised algorithm is outputted. You can give this model new data that it has not seen before and get a prediction on what the pricing of the car would be. There are various types of supervised learning algorithms.

Types of Supervised Learning Algorithms

Regression

Regression algorithms predict continuous variables such as pricing or temperature.

Examples of Supervised Learning Algorithms for Regression

- Linear Regression
- Random Forest Regression
- Decision Tree Regression
- Support Vector Regression

Classification

Classification algorithms predict discrete variables such as

- Yes or No
- 0 and 1
- Red, Purple, Blue
- True or False

The prediction is categorized into a distinct set of choices.

Examples of Supervised Learning Algorithms for Classification

- Support Vector Classification
- K-Nearest Neighbors (KNN)

- Logistic Regression
- Decision Tree Classification
- Random Forest Classification

Use Cases of Supervised Learning

- Time Series Market Analysis
- Customer Analytics
- Human Resource Allocation

2. Unsupervised Learning

Machine learning algorithms take unlabeled datasets and find similarities and differences in the information without knowing the data beforehand. For example, if you are looking at a dataset of used Lamborghinis, the algorithm can automatically group the cars based on similarities like model or color. This way, clusters in the data can be identified, such as by grouping cars with models together.

Types of Unsupervised Learning Algorithms

Clustering

Clustering algorithms group your data by finding patterns in your dataset. The items with the most similarities will be placed in the same group.

Some Unsupervised Learning Algorithms Using Clustering

- Gaussian Mixture
- K-Means
- Isolation Forest

Association Rule

This finds the relationship between data points within your dataset. It finds the set of items that happen together. For instance, people who buy flowers often tend to buy chocolate.

Some Unsupervised Learning Algorithms Using Association Rule

- Apriori
- Eclat

Some Use Cases of Unsupervised Learning

- Figuring out customer personas
- Anomaly detection
- Object recognition

3. Semi-supervised Learning

Machine learning algorithms take datasets with a mixture of labeled and unlabeled data as input. Typically, there is a small amount of labeled data and a large amount of unlabeled data.

How It Works

1. Train the model with labeled data.
2. Train the partially trained model in step 1 with the unlabeled data to predict outputs, which are considered pseudo labels since it may not be accurate.
3. Link the predicted output from the model of the labeled data with the predicted output from the model of the unlabeled data.

4. Link the data inputs of the labeled training data to the data inputs of the unlabeled training data.

5. Train the model again to get more accurate predictions.

Some Use Cases of Semi-supervised Learning

- Label and Rank Web Pages in Search Results
- Image and Audio Analysis

4. Reinforcement Learning

Machine learning algorithms are trained to make a sequence of decisions. There is an agent (the algorithm) and rewards with many obstacles throughout. Unlike supervised and unsupervised learning, there is no external training dataset provided beforehand. The agent generates its own training data by taking action. It learns from trial and error to solve a problem. The agents are rewarded or penalized depending on the decision it makes. The goal is to maximize the number of rewards.

Some Use Cases of Reinforcement Learning

- Video Games
- Autonomous Vehicles
- Customized Training Systems

Deep Learning Models

Deep learning is a subset of machine learning. It focuses on neural networks with many layers (hence "deep") that attempt to mimic the human brain's structure and function. These layers are composed of interconnected nodes, also known as neurons, that work together to process information and make decisions.

Neural Networks: An AI technique that enables computers to process data by mimicking the structure and function of the human brain. This machine learning approach, known as deep learning, involves layers of interconnected nodes or neurons that resemble the brain's neural pathways. It forms an adaptive system that allows computers to learn from errors and improve over time.

Deep learning models have many layers. The most basic neural network has three layers. A simple neural network can be seen in Figure 5-2.

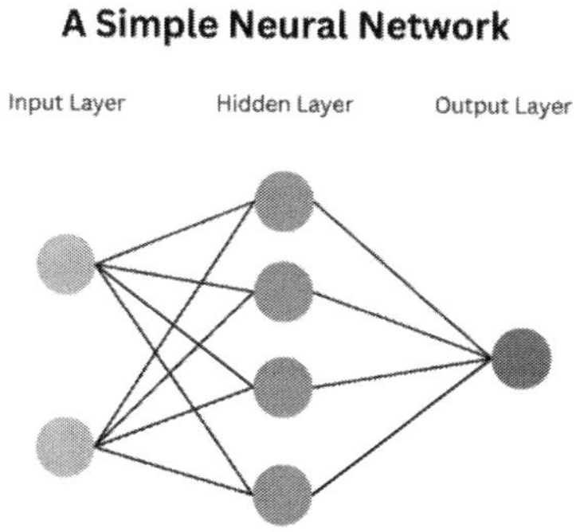

Figure 5-2. *A visual representation of a simple neural network*

1. Input Layer: The first layer that receives the data
2. Hidden Layer: Intermediate layer that processes input from the input layer
3. Output Layer: The final layer that produces the prediction or decision

Deep learning networks have several hidden layers with several nodes linked together, as seen in Figure 5-3.

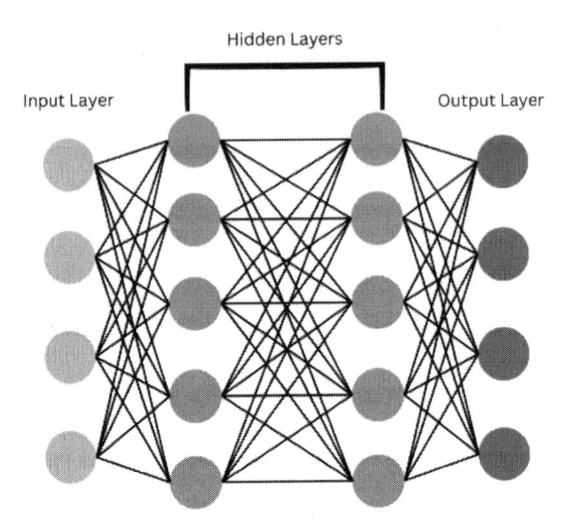

Figure 5-3. *A visual representation of a deep neural network*

Unlike traditional machine learning methods, where data scientists have to choose and create the important features that help the model learn from the data, deep learning methods can automatically figure out these important features independently. It discovers patterns and details directly from the raw data without needing extra help. This makes them really good at understanding complex information. However, it doesn't hurt to do some initial feature selection and engineering beforehand to give the model a break if you have some domain knowledge on the tasks and data at hand! Deep learning models also require more computational power and more data to train it.

Key Concepts and Terminology

- Feed Forward: A feed forward neural network is a basic type of neural network characterized by its straightforward information flow. Data travels in only one direction, from input nodes to hidden nodes and ultimately to the output nodes. Each node processes the input and passes it to the next layer.
- Backpropagation: Backpropagation is a learning method for neural networks. It's like teaching the network to learn from its mistakes. It helps improve neural networks by figuring out how far off it was and then makes adjustments to its settings to get closer to the correct answer.
- Recurrent Neural Network (RNN): RNN is a type of neural network that is trained to handle data that comes in a sequence, like a series of events over time. This allows the RNN to make predictions or decisions based on the order of the inputs. For example, an RNN may be used to predict how much a stock price will change by looking at past prices and market trends.

  RNNs are unique because they have "memory." This means they use information from earlier inputs to help understand the current input to help produce the correct output. Unlike other types of neural networks, which treat each input separately, RNNs rely on the order of the sequence.

Imagine you are trying to bake a cake. The order of the steps is essential. You can't bake the cake without mixing the ingredients in order. Similarly, an RNN remembers the sequences of steps and uses that to predict what comes next.

- Long Short-Term Memory (LSTM): LSTMs are special types of RNNs designed to fix a problem that regular RNNs have. Regular RNNs can struggle to remember details when they are dealing with long sequences of information. This makes it difficult to connect things that happened a while ago to what's happening now. LSTMs solve this issue because they are built to remember information over long periods. In fact, holding onto important details for a long period of time is what they are really good at!
- Convolutional Neural Network (CNN), a.k.a. ConvNets: CNNs are typically used for object detection and image classification tasks.
- Generative Adversarial Network (GAN): GANs are made up of two neural networks: one is called the generator, and the other is called the discriminator. The generator's job is to create new samples that resemble the training dataset, while the discriminator's job is to tell the difference between the real samples and the ones the generator makes. The term "adversarial" comes from how the two networks are set up to compete. A GAN could be used to create new images based on a collection of existing images or to generate new music by learning from a database of songs. GANs can be used to create realistic images from text

descriptions or to improve existing photos. They can also be used to make new synthetic data that has the same characteristics as real data.

- Transformers: Transformers are a type of neural network that changes an input sequence, like a sentence, into an output sequence by understanding the context and relationships between the parts of the sequence.

  Unlike other neural networks like RNNs and CNNs, transformers are faster and more efficient because they process sequences all at once, rather than one step at a time. This means they need less training time and can work more quickly.

  Transformers have two important features that make them powerful for tasks like predicting text:

  i. Positional Encoding: Instead of just reading words in the order they appear, transformers assign a unique number to each word. This helps the model understand the order of the words in a sentence, which is important for making sense of the meaning.

  ii. Self-attention: This feature allows the model to focus on the relationships between all the words in a sentence at the same time. It learns which words are important to each other, helping the model predict what comes next in a sequence. For example, transformers can "learn" grammar rules by figuring out how words usually fit together.

Overall, transformers are powerful tools for understanding and predicting sequences, like sentences, in a smart and efficient way.

- Large Language Model (LLM): A deep learning model that is only able to understand and generate text data.
- Multimodal Models (MM): A deep learning model that is capable of handling different modalities. Examples of this can be text, audio, or images. A model being multimodal can mean one or more of the following:

  i. The modalities of the input and output are different (i.e., image - to - text).

  ii. The inputs to the model are multimodal (i.e., the model can accept both text and images).

  iii. The outputs of the model are multimodal (i.e., the model can generate both text and images).

- Large Multimodal Model (LMM): A deep learning model that is a combination of a LLM and MM.
- Retrieval Augmented Generation (RAG): RAG is a technique used to make machine learning models more accurate by pulling in facts from outside sources. LLMs and LMMs are powerful, and they can do things like answer questions, extract texts from documents, or translate languages, but they have some limits. For example, they can only use the information they were trained on, which might be outdated or incomplete. Sometimes, they might even give wrong answers if they don't know something.

RAG helps solve these problems by allowing the model to check a reliable source of information before giving an answer. This allows the model to provide more accurate and up-to-date responses without needing to be retrained or fine-tuned.

- Generative Artificial Intelligence (GEN AI): A deep learning model that has the ability to create new content based on a user's request or prompt.

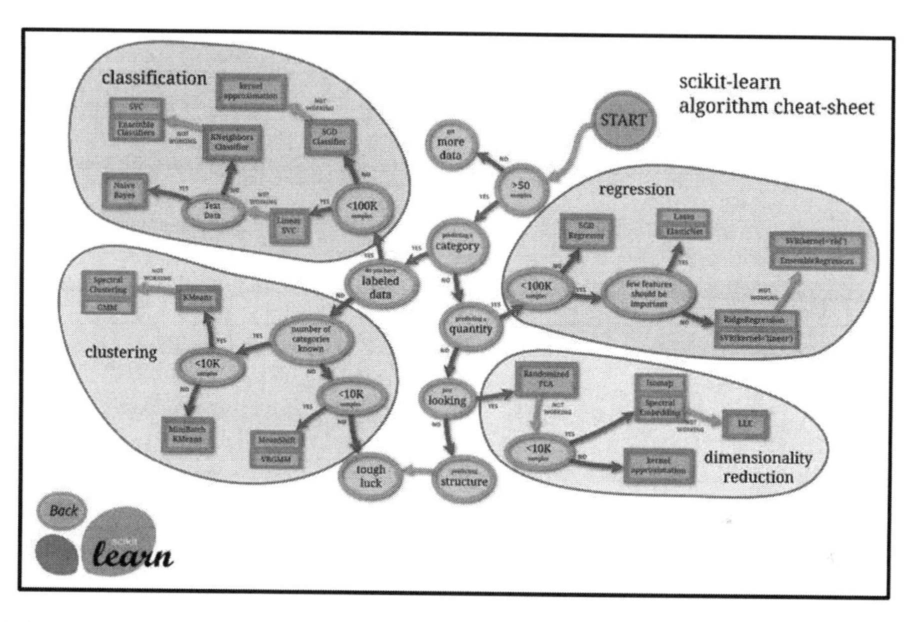

Figure 5-4. *A guide from the scikit-learn website to help in the selection of machine learning algorithms, such as classification, clustering, and regression, based on dataset size, number of features, and types of issues addressed ([scikit-learn algorithm cheat-sheet],n.d)*

Machine learning models come in various forms, each with strengths and weaknesses. The choice of model depends on the specific problem, the nature of the data, and the desired outcome.

Importance of Machine Learning Models

Machine learning models have the potential to be crucial across various sectors.

Healthcare

In healthcare, ML models can be instrumental in enhancing diagnostic accuracy, personalizing treatment plans, predicting patient outcomes, and assisting in educational outreach. For instance, deep learning models can be utilized to analyze medical images and detect anomalies in your body, such as fractures and pathologies. Models can assist in playing a crucial role in identifying patients at risk of developing chronic conditions such as diabetes and heart disease, enabling early interventions that can improve patient outcomes in life. Models can also play a role in accelerating the process of extracting patient information from handwritten and scanned documents.

Finance

The financial sector can use machine learning models in many ways, such as fraud detection and forecasting. Fraud detection models analyze patterns to identify fraudulent behavior, protecting financial institutions and their customers. In forecasting, a machine learning model can analyze market trends and predict potential risks, helping financial institutions make informed decisions.

Retail

Machine learning can enhance customer experience in retail through recommendation systems, forecasting, and inventory management. By analyzing customer behavior and purchase history, recommendation engines can suggest products that align with individual preferences, thereby increasing sales and customer satisfaction. Forecasting models can predict future sales and trends, enabling retailers to optimize their inventory and reduce costs associated with overstocking.

Transportation

Autonomous vehicles and traffic management systems exemplify the ways AI can have a transformative impact in the transportation sector. Self-driving cars leverage machine learning models for navigation, object detection, and decision-making. This aims to improve safety and efficiency on the roads. Traffic management systems can use predictive models to optimize traffic flow, which can reduce congestion and minimize travel times by analyzing real-time data. The model can be used for incident detection and management. For example, it can be used to detect accidents, wrong-way drivers, and irregular driving like drunk drivers swerving. This information can be used to alert the authorities.

Marketing

Machine learning models can be used to analyze consumer data to target advertisements, segment markets, and predict customer churn. This can help businesses tailor their marketing strategies to individual preferences and behaviors. This can lead to more effective campaigns and higher conversion rates. Companies can proactively implement retention strategies to maintain a loyal customer base by understanding customer lifetime value and identifying potential churners.

Machine Learning Models' Tools and Techniques

Programming Languages

Python is one of the most popular languages for machine learning. It is known for its readability and extensive libraries, such as TensorFlow, Keras, scikit-learn, PyTorch, and Hugging Face Transformers. Python's simplicity and versatility make it an ideal choice for beginners and experienced practitioners. The language supports rapid prototyping and development, enabling data scientists and machine learning engineers to experiment with different algorithms and techniques quickly and efficiently.

Frameworks and Libraries

TensorFlow

TensorFlow is an open source framework developed by Google, used for building and deploying ML models. It provides a flexible and comprehensive ecosystem for machine learning development, including tools for model training, evaluation, and deployment.

PyTorch

Developed by Meta, PyTorch is known for its ease of use in research and development. PyTorch is a favorite among researchers and developers for prototyping and experimenting with new ideas.

Keras

Keras is a high-level neural network API written in Python and capable of running on top of TensorFlow and PyTorch. It simplifies the process of building and training deep learning models by providing a user-friendly

API and good documentation. Keras is beneficial for beginners and practitioners who want to develop and experiment with neural networks quickly.

scikit-Learn

scikit-learn is a library for classical ML algorithms like regression, classification, and clustering. It provides simple and efficient tools for data analysis and modeling. It is a popular choice for educational purposes and real-world applications. scikit-learn's well-documented API and extensive collection of algorithms make it a go-to library for implementing standard machine learning tasks.

Transformers

Hugging Face Transformer is an open source library that simplifies the use of state-of-the-art machine learning models. It provides a wide range of pre-trained models for tasks such as image classification, named entity recognition, and question answering, among others. Built on top of PyTorch and TensorFlow, the library offers an accessible interface for leveraging these powerful models, enabling developers and researchers to integrate advanced machine learning capabilities into their projects with ease.

Integrated Development Environments

Jupyter Notebook

Jupyter Notebook is an open source web application that allows for interactive coding, visualization, and sharing of ML models. It supports multiple programming languages and integrates seamlessly with various data science libraries. Jupyter Notebooks enable data scientists to document their workflow, visualize results, and collaborate with others in a reproducible manner.

Google Colab

Google Colab is a free cloud service that supports Jupyter Notebooks and provides free access to GPUs, making it suitable for ML experiments. It offers a convenient platform for developing and testing machine learning models without the need for complex setup or hardware resources. Google Colab's integration with Google Drive allows for easy storage and sharing of notebooks and data.

Other IDEs that can be used for machine learning development include Visual Studio Code and PyCharm.

EXAMPLE WITH CODE

In this example, we will use a machine learning model to predict salaries based on the number of years of work experience.

Import required libraries.

```
import pandas as pd
from sklearn.model_selection import train_test_split
from sklearn.linear_model import LinearRegression
import matplotlib.pyplot as plt
import pickle
```

Load the dataset. This dataset can be found on Kaggle: `https://www.kaggle.com/datasets/abhishek14398/salary-dataset-simple-linear-regression`.

```
examp1_data = pd.read_csv("salary_dataset.csv")
examp1_data.head(5)
```

	YearsExperience	Salary
0	1.2	39344
1	1.4	46206
2	1.6	37732
3	2.1	43526
4	2.3	39892

Figure 5-5. *A sample of the data within the dataset we are using for this example. This is the first rows in the dataset*

We will use years of experience as the input to the model and salary as what we want our model to output.

```
x = examp1_data[["YearsExperience"]]
y = examp1_data[["Salary"]]
```

We should first split the dataset to ensure we have enough data to train the model and enough to test the model. Typically, the data is split with 80 percent of the data used for training and 20 percent of the data used for testing.

We will use `x_train` and `y_train` for training and `x_test` and `y_test` for testing the model.

`random_state` is a parameter set to ensure the reproducibility of the split in data. So every time you run this line of code, the same data will be set to `x_train, x_test, y_train, y_test` instead of random numbers.

```
x_train, x_test, y_train, y_test = train_test_split(x, y,
train_size=.8, random_state=42)
```

```
model = LinearRegression()
model.fit(x_train, y_train)
```

Let's use our model to predict salaries using the years of experience data the model has not seen. This data was saved in the variable `x_test`. Next, let's save those results in `lr_predictions` and plot them.

```
lr_predictions = model.predict(x_test)
```

Now, let's plot the data.

```
plt.scatter(x_test, y_test)
plt.plot(x_test, lr_predictions)
plt.xlabel("Years")
plt.ylabel("Salary")
plt.title("Title: Predicted Salary Based on Years of
Experience")
```

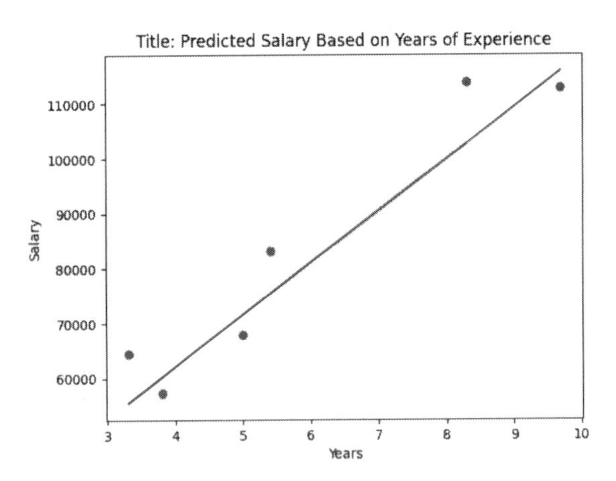

***Figure 5-6.** A graph displaying the model's predictions of salary in the next ten years*

Now, let's save the model so that we can use it later. We save it in a pickle file and name it `lr_salary_determinator_model.pkl`

```
lr_model_pickle_file = "lr_salary_determinator_model.pkl"
with open(lr_model_pickle_file, "wb") as file:
  pickle.dump(model, file)
```

Now, let's load our pickle file and use the model with the same data in x_test we used previously.

```
with open("lr_salary_determinator_model.pkl","rb") as file:
  model = pickle.load(file)
predictions = model.predict(x_test)
```

In this example, we will use a deep learning model to identify animals in an image.

Import required libraries.

```
from transformers import pipeline
from PIL import Image
```

In this example, we will use an image of an animal.

```
image = Image.open("ostrich.jpg")
```

We will use a model from Hugging Face called google/vit-base-patch16-224.

```
model_name= "google/vit-base-patch16-224"
image_classifier = pipeline("image-classification",
model=model_name)
```

Here, we input the image into the model and wait for its predictions.

```
image_classification_prediction = image_classifier(image)
```

Below are the outputs from the model. It outputted five predictions in order of how confident it is. The first is an ostrich with .99 confidence. The confidence scores for models typically range from 0 to 1 or 0 to 100%.

```
[{'label': 'ostrich, Struthio camelus', 'score':
  0.9992263317108154},
 {'label': 'bustard', 'score': 9.642778604757041e-05},
```

```
{'label': 'black stork, Ciconia nigra', 'score':
 4.994613482267596e-05},
{'label': 'zebra', 'score': 1.568139487062581e-05},
{'label': 'flamingo', 'score': 1.4048618140805047e-05}]
```

Feature Selection and Engineering

Feature selection is about picking out the most important pieces of information (features) from a dataset and getting rid of the unnecessary stuff that could confuse the model. On the other hand, feature engineering involves creating new pieces of data that aren't initially in the dataset but could help the model make better predictions.

When we train a machine learning model, we start off with a lot of data, but not all of it is useful. Some data might be noisy, irrelevant, or just extra information that just slows down the model's training. By selecting only the best features, we can help the model work faster and make better predictions.

For example, imagine you are building a machine learning model to predict which students will do well in a math competition. You have a dataset that includes the students' grades in math, the number of math books they own, and their favorite TV shows. The grades and the number of math books might be helpful, but their favorite TV shows probably won't help predict how well they will do in the competition. So you would remove the TV show data and focus on the more relevant features to build the model. However, you could create a new feature through feature engineering. For example, you could take their favorite TV shows and use that to create a feature that identifies which TV shows have a strong math theme and create a binary feature (e.g., Math-Focused-Show = 1 or 0) to see if that plays a role in how good a student will be in the competition.

Pre-trained vs. Fine-Tuned Models

Pre-trained Model: A model already trained on large amounts of data to perform a task. It can be used as it is or tailored to fit different needs across various industries and use cases using methods like fine-tuning. The model in our second code example is a pre-trained model.

Fine-Tuned Model: Fine-tuning is the process of taking a pre-trained model and adjusting it to work better for a specific task by using a smaller, more focused dataset. This helps the model refine its skills and perform better on the particular task at hand.

Prompt Engineering

A prompt is a text that asks a machine learning model to do something specific, like answering a question or completing a task. Prompt engineering is carefully crafting and adjusting these prompts to get the model to give the exact responses you want. It's like the way you give instructions or ask a person questions. In the world of AI, where the models learn from a lot of data, the right prompt can make a big difference in whether the model understands your request or gets it wrong.

Conclusion

Machine learning models have transformed various industries by automating complex tasks, uncovering insights from vast datasets, and enabling data-driven decision-making. The tools and techniques discussed in this chapter provide a comprehensive foundation for building and deploying ML models. Through many examples, we have demonstrated the practical applications and effectiveness of these models in solving real-world problems. As machine learning continues to evolve, it will unlock new opportunities and drive innovation across diverse domains.

Machine learning models have revolutionized numerous fields by providing powerful tools for data analysis, prediction, and automation. By leveraging advanced techniques and robust tools, these models can solve complex problems and drive significant improvements across industries. Understanding the importance, tools, techniques, and potential applications of ML models is essential for harnessing their full potential.

CHAPTER 6

Evaluating Models

Evaluating machine learning models is like checking to see if a recipe turned out the way you wanted. Imagine you are baking cookies and want them to be my favorite type, soft and chewy! After baking, you take a bite to see if they are just right. If they're too hard or soft, you know something went wrong. In the same way, when we build a machine learning model, we need to test it to ensure it is doing what we want. This testing, or evaluation, helps us determine if the model is making accurate predictions or if we need to adjust the "ingredients" to get better results.

Without proper evaluation, we might end up trusting a model that doesn't work well, just like serving cookies that do not taste good. This could lead to bad decisions, especially if the model is used in important areas like healthcare. So evaluating models helps ensure they are reliable and can be trusted to make the right decisions. Various metrics can be used to evaluate machine learning models, depending on their task.

Introduction to Evaluating Models

Building models is just the beginning in the world of machine learning. Evaluating these models is critical to ensuring their effectiveness and reliability in real-world applications. Model evaluation involves using various metrics and techniques to assess how well a machine learning model performs on a given dataset.

M. Attobrah, *Essential Data Analytics, Data Science, and AI*,
https://doi.org/10.1007/979-8-8688-1070-1_6

Importance of Evaluating Models

Evaluating machine learning models is vital for several reasons:

1. Performance Assessment

   It helps understand how well a model performs on unseen data, ensuring that the model does not just memorize the training data but generalizes well to new, unseen information.

2. Model Comparison

   Evaluation metrics allow for the comparison of different models or algorithms to select the one that performs best for a specific task.

3. Identifying Overfitting and Underfitting

   Overfitting happens when a machine learning model learns too much from its training data, almost memorizing it. This means it does well with familiar data but struggles when given new information. It's like a student who memorizes all the answers for a test but can't answer new questions.

   Some reasons why overfitting can happen:

   - The training process is too long.
   - The training data is not diverse enough.
   - The model is too complex for the data.

   Underfitting happens when a machine learning model does not learn enough from the data, leading to poor performance on both the training and testing data. It's like a student who only skimmed the textbook and missed key concepts.

Some reasons why underfitting can happen:

- The training is not long enough.
- The model is too simple for the data.
- The data has too many unnecessary details.

Finding the right balance between overfitting and underfitting is key to making sure a model can handle both familiar and new data.

4. Hyperparameter Tuning

   Let's say you wanted to teach a robot how to identify cats. Hyperparameters are like the rules you set for the robot to learn. They tell the robot how fast to learn (learning rate) and how many times to look at the same pictures (epochs).

   This is different from model parameters. Model parameters are what the robot learns from the pictures. These are like the specific details the robot remembers about cats, like the shape of their ears and their fur color.

   Hyperparameters help the robot learn, while the model parameters are what the robot actually learns.

   Another way to think about it is baking a cake. Hyperparameters are like the recipe you follow. They tell you how much of each to use (e.g., two cups of flour), how long to bake the cake (e.g., 30 minutes), and the temperature of the oven (e.g., 350 °F). These are choices you make before baking the cake.

The actual cake itself would be the model parameters. They are the result of following the recipe. The cake's specific flavors, texture, and appearance depend on how well you followed the recipe and the quality of your ingredients.

So hyperparameters guide the baking process, while model parameters represent the final product. Examples of hyperparameters are

- Learning Rate: Determines how quickly the algorithm updates its predictions as it learns
- Weight Decay: Gradual reduction of the learning rate over time to help the model learn more effectively
- Epochs: The number of times the model goes through the entire training data set during the learning process

5. Deployment Readiness

   Thorough evaluation ensures that a model meets the necessary accuracy and reliability standards before it is deployed in a production environment.

Tools and Techniques When Evaluating Models'

Model evaluation employs a variety of tools and techniques, each suited for different types of problems and data. Below, we discuss some of the most common methods and metrics used in the evaluation of machine learning models. I discussed some of these techniques in a few blogs and will add the definitions and discussions here.

1. Train-Test Split

   The train-test split is a fundamental method where the dataset is divided into two parts: the training set and the testing set. The model is trained on the training set and evaluated on the testing set to assess its performance.

2. Cross-Validation

   Cross-validation involves dividing the dataset into multiple folds and training the model on different folds while evaluating it on the remaining fold(s). This helps estimate how the model will perform and prevents the model from overfitting. This technique is used to help identify a model's hyperparameters.

   There are various types of cross-validation techniques. In this book, we will discuss k-fold cross-validation.

   K-fold cross-validation is when the dataset is split into K number of groups. These groups are used to train and test the model, K number of times. The portions used to train and test the model are changed once during each iteration until you reach the K-ith iteration.

   How It Works

   1. Select a value for K. A commonly used value is 10.
   2. Split your dataset into K number of groups.
   3. K-1 groups will be used as training data to train your model, and the rest will be used as testing data to evaluate your model.

4. Train your model on the training dataset and evaluate your model on the testing dataset.

5. Evaluate your model using evaluation metrics, such RMSE, MAE, accuracy, etc. These metrics will be discussed further in this chapter. After evaluating the model, save your evaluation score.

6. Repeat steps 1–5 K times. However, in the next round, use a different section as your testing dataset.

7. Finally, take the average of the evaluation scores.

In Figure 6-1, I show a visual representation of five-fold cross-validation. Luckily, we do not have to do this manually. There are tools available for this like Python packages, which can help minimize mistakes when dividing the dataset.

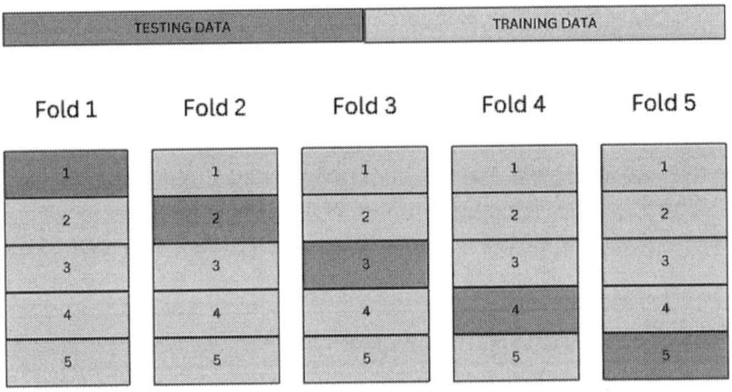

Figure 6-1. *A visual representation of five-fold cross-validation*

3. Confusion Matrix

A confusion matrix provides a detailed breakdown of the model's predictions, showing the true positives, true negatives, false positives, and false negatives. It visually represents how well the model is at making predictions by displaying the results in a table. The rows of the table represent the expected output, and the columns represent the predicted output from the model.

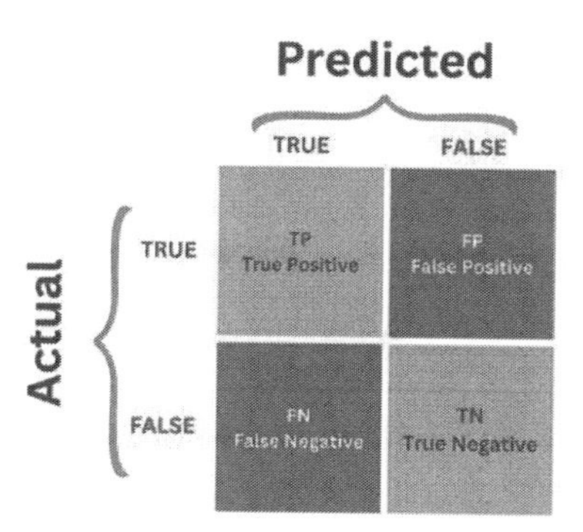

Figure 6-2. *A visual representation of a confusion matrix*

Let's say you wanted to create a model that predicts whether a car is a Lamborghini. A **positive** case in this dataset means that the car is a Lamborghini, and a **negative** case is any other car. There are four terms to keep in mind when assessing the model using this.

- True Positives (TP): Data points within the dataset that are positive AND the predictions from the model are also positive
- True Negatives (TN): Data points within the dataset that are negative AND the predictions from the model are also negative
- False Negatives (FN): Data points within the dataset that are positive BUT the predictions from the model are negative

- False Positives (FP): Data points within the dataset that are negative BUT the predictions from the model are positive

4. Metrics

Different metrics are used depending on the type of machine learning problem. You can use these metrics to evaluate the model and compare performances between multiple models. Below is a list of some metrics to consider:

Classification Metrics

Accuracy

Using the explanation from the confusion matrix above, accuracy is the number of correct predictions divided by the total number of predictions. This number ranges between 0 and 1, where zero means the model did not get any predictions correct and one means all the predictions were correct. Business stakeholders love this metric because of its simplicity of use.

Let's say you wanted your model to determine if a flower basket contains roses. Using your validation dataset, the accuracy of your model would be determined by how many items it correctly identified as roses and how many items it correctly identified as not being roses, divided by the total number of flowers.

This metric is only effective if your dataset is balanced. A balanced dataset means that every group is equally represented. If one group has way more examples than others, the model might always predict the bigger groups, even though figuring out the smaller groups is just as important. This can lead to high accuracy but poor performance in recognizing the less common groups, which can be problematic in real-world applications where identifying those minority groups is crucial.

Precision

Out of all the positive predictions that the model outputted, how many were actually positive. This measures how many items the model correctly identified divided by the total number of items it correctly identified and the total number it incorrectly identified.

This keeps track of items the model incorrectly identified. Using our flower example, it keeps track of how many of our classified roses were actually sunflowers.

Recall

Out of all the actual positive predictions, how many did the model predict correctly. This measure is the number of times the model correctly identified divided by the total number of items present. This takes into consideration the number of items that the model ignored. It keeps track of the total number of roses we may have missed in the flower dataset.

F1 Score

The F1 score takes into account precision and recall. It measures how many items were misclassified by the model and how many positives were completely missed by the model.

ROC-AUC

- Receiver Operating Characteristic (ROC): A graph that shows a plot of the True Positive Rate vs. the False Positive Rate at different thresholds

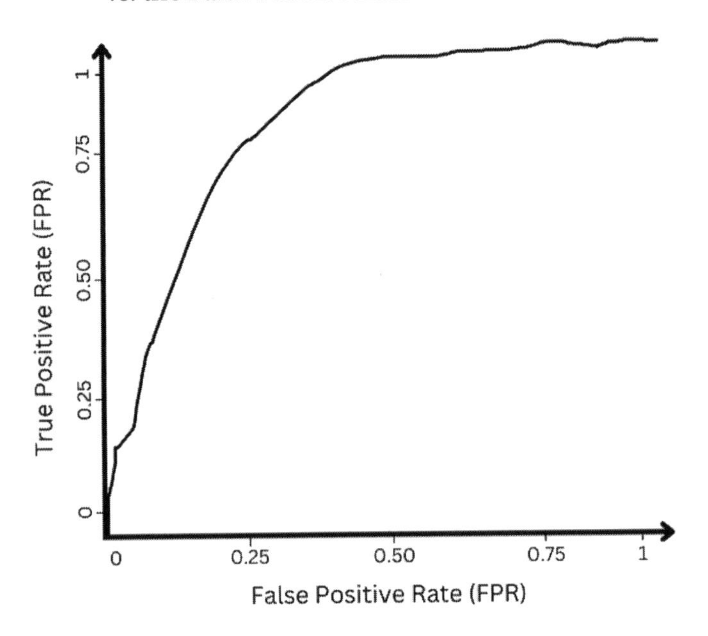

Figure 6-3. *An example of the ROC curve*

- True Positive Rate (TPR): Probability that data points within the validation dataset were positive AND the predictions from the model also outputted positive

- False Positive Rate (FPR): Probability that data points within the validation dataset were negative BUT the predictions from the model outputted positive

- Area Under the Curve (AUC): Measures the area under the ROC curve. The higher the AUC, the better the model is predicting.

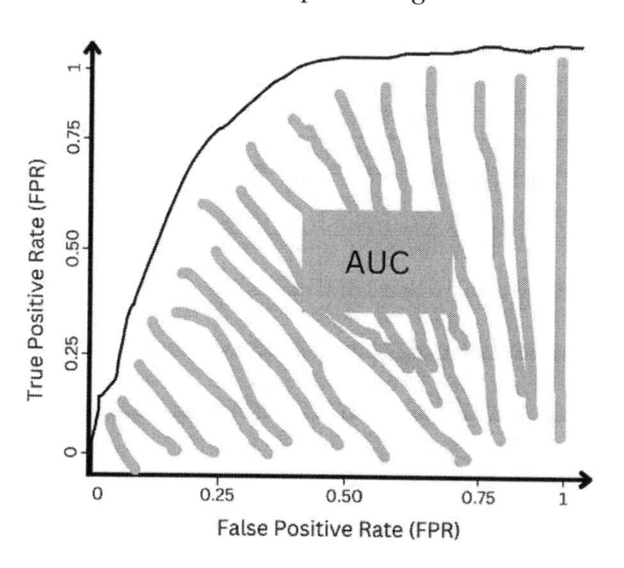

Figure 6-4. *An example of AUC*

Sometimes AUC and ROC are referred together as AUROC (Area Under the Receiver Operating Characteristic) curve.

Regression Metrics

Mean Squared Error (MSE)

MSE measures how close the data points predicted by the model are to the data points in your validation dataset. It is the average of the square of the residuals. Residuals are the difference between the data points predicted by a model and the data points in the validation dataset. Residuals are also known as the errors. MSE ranges from 0 to infinity. The closer the MSE is to zero, the better the model is performing. In other words, MSE should be as low as possible.

Root Mean Squared Error (RMSE)

RMSE is the square root of MSE. It measures how close the data points a model predicts are to the data points in your validation dataset. Similarly to MSE, the values range from 0 to infinity. The closer the RMSE is to 0, the better the model is performing. There is a higher preference for RMSE over MSE because the units are not squared, so it's easier to interpret the performance of the model.

Mean Absolute Error (MAE)

MAE measures the average difference between the data points predicted by a model and the data points in your validation dataset. It is the average absolute error between the predicted values from the model and the validation dataset. The closer the MAE is to 0, the better the model is performing.

R Squared

R squared is also called the coefficient of determination. It measures how close the actual data within your validation dataset is to the fitted line, also known as the regression line.

The regression line is a straight line that represents how the target variable changes as the explanatory variables change. The target variable is what you want your model to predict, and the explanatory variables are the variables chosen as input to help your model make the predictions.

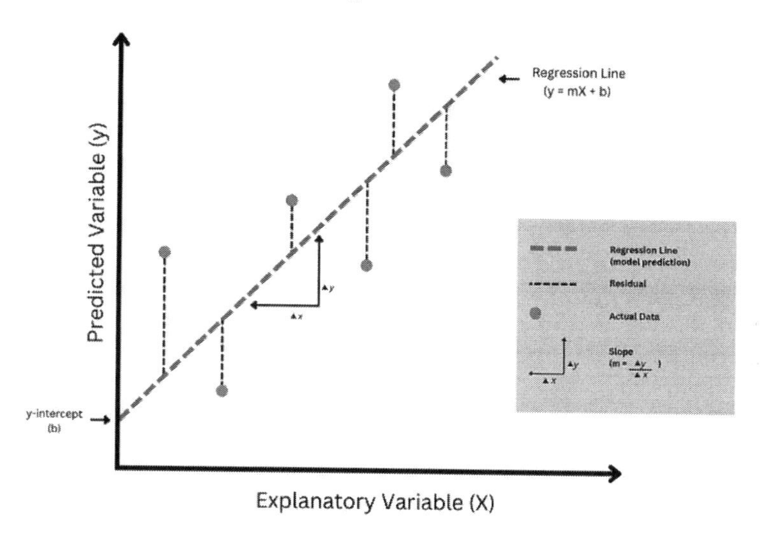

Figure 6-5. *A visual representation on R squared*

R squared also measures how close the data points predicted by a model are from the data points in your validation dataset. It is best to use R squared for linear regression machine learning models. The values range from 0 to 1. The closer the value is to 1, the better the model is performing.

Let's go back to our example in Chapter 5, where we used a machine learning model to predict salaries. In the last chapter, we created a model without evaluating it. Now that we have learned some evaluation metrics, let's evaluate the model using R squared.

Import required libraries.

```
import pandas as pd
from sklearn.model_selection import train_test_split
from sklearn.linear_model import LinearRegression
import matplotlib.pyplot as plt
```

Load the dataset. This dataset can be found on Kaggle: https://www.kaggle.com/datasets/abhishek14398/salary-dataset-simple-linear-regression.

```
examp1_data = pd.read_csv("salary_dataset.csv")
examp1_data.head(5)
```

	YearsExperience	Salary
0	1.2	39344
1	1.4	46206
2	1.6	37732
3	2.1	43526
4	2.3	39892

Figure 6-6. *A sample of the data within the dataset we are using for this example*

We will use years of experience as the input to the model and salary as what we want our model to output.

```
x = examp1_data[["YearsExperience"]]
y = examp1_data[["Salary"]]
```

We should first split the dataset to ensure we have enough data to train the model and enough to test the model. Typically, the data is split with 80 percent of the data used for training and 20 percent of the data used for testing.

We will use `x_train` and `y_train` for training and `x_test` and `y_test` for testing the model.

`random_state` is a parameter set to ensure the reproducibility of the split in data. So every time you run this line of code, the same data will be set to `x_train, x_test, y_train, y_test` instead of random numbers within the dataset each time you run.

```
x_train, x_test, y_train, y_test = train_test_
split(x, y, train_size=.8, random_state=42)
```

Let's train a Linear Regression model using our data.

```
model = LinearRegression()
model.fit(x_train, y_train)
```

Now, let's use our trained model to predict salaries using the years of experience data the model has not seen. This data was saved in the variable `x_test`. Next, let's save those results in `lr_predictions` and plot them.

```
lr_predictions = model.predict(x_test)
```

Finally, let's plot the data.

```
plt.scatter(x_test, y_test)
plt.plot(x_test, lr_predictions)
plt.xlabel("Years")
plt.ylabel("Salary")
plt.title("Title: Predicted Salary Based on Years of
Experience")
```

As Figure 6-7 shows, our regression line from the model and the actual data, which are the dots, are close.

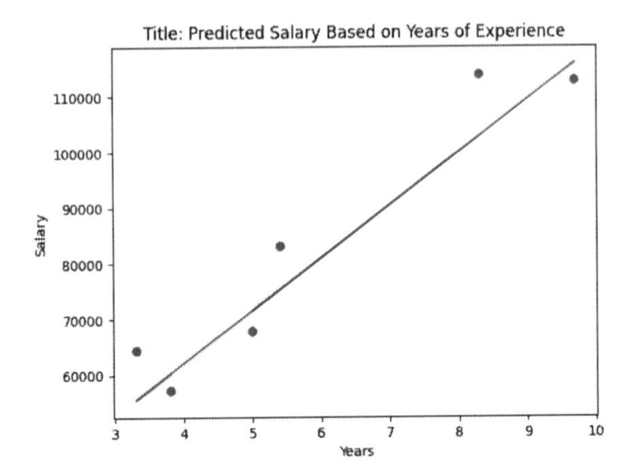

***Figure 6-7.** A graph displaying the model's predictions of salary in the next ten years*

Let's calculate R squared using `r2_score` from `scikit-learn`.

```
r2_score(y_test, lr_predictions)
```

Our result is `0.9024461774180497`, which is very close to 1.

5. Learning Curves

   Learning curves plot the model's performance on the training and evaluation sets over successive training interactions, helping in diagnosing whether the model is overfitting or underfitting.

Examples with Code

EXAMPLE WITH CODE

In this example, we will use a machine learning model to predict the quality of wine. There are 1600 rows in this dataset. The quality of wine ranges from 3 to 8, where 8 is good and 3 is bad.

Import required libraries.

```
import pandas as pd
from sklearn.ensemble import RandomForestClassifier
from sklearn.model_selection import train_test_split,
GridSearchCV
from sklearn.metrics import accuracy_score,
classification_report
import matplotlib.pyplot as plt
```

Load the dataset. This dataset can be found on Kaggle:

```
https://www.kaggle.com/datasets/uciml/red-wine-quality-
cortez-et-al-2009/data

wine_data = pd.read_csv("wine-quality-data.csv")
wine_data.head(5)
```

We will use `quality` of the wine as what we want our model to output and the other columns as input to the model to help make its predictions. That is the last column seen in Figure 6-8.

```
x = wine_data.drop(["quality"], axis =1)
y = wine_data["quality"]
```

	fixed acidity	volatile acidity	citric acid	residual sugar	chlorides	free sulfur dioxide	total sulfur dioxide	density	pH	sulphates	alcohol	quality
0	7.4	0.70	0.00	1.9	0.076	11.0	34.0	0.9978	3.51	0.56	9.4	5
1	7.8	0.88	0.00	2.6	0.098	25.0	67.0	0.9968	3.20	0.68	9.8	5
2	7.8	0.76	0.04	2.3	0.092	15.0	54.0	0.9970	3.26	0.65	9.8	5
3	11.2	0.28	0.56	1.9	0.075	17.0	60.0	0.9980	3.16	0.58	9.8	6
4	7.4	0.70	0.00	1.9	0.076	11.0	34.0	0.9978	3.51	0.56	9.4	5

Figure 6-8. *A sample of the data within the dataset we are using for this example. Here are the first five rows*

For this example, we will use a Random Forest Classification algorithm to create a model to make the wine quality predictions.

```
model = RandomForestClassifier(random_state =42)
```

Let's use `GridSearchCV` to help us pick hyperparameters for training. `GridSearchCV` is a tool used to automate the process of finding the best combination of hyperparameters for a machine learning model.

```
hyper_params = {'criterion':['gini', 'entropy','log_loss'],
                'max_depth':[5, 25, 50,100,200,300, None],
                'n_estimators':[50,100,200, 500]}
grid = GridSearchCV(estimator = model, param_grid = hyper_
params, scoring="accuracy", cv =10)
grid.fit(x, y)
```

Let's see what it came up with.

```
print(grid.best_params_)
print(grid.best_score_)
```

Our print statements outputted:

```
{'criterion': 'gini', 'max_depth': 5, 'n_estimators': 200}
  0.5922680817610062
```

So out of the list of options we gave `GridSearchCV`, the best parameters for the model should have the criterion set to `'gini'`, max_depth set to `5`, and n_estimators set to `200`. This will give us an estimated accuracy score of `0.59`.

We will now split the dataset to ensure we have enough data to train the model and enough to test the model. Typically, the data is split with 80 percent of the data used for training and 20 percent of the data used for testing.

We will use `x_train` and `y_train` for training and `x_test` and `y_test` for testing the model.

`random_state` is a parameter set to ensure the reproducibility of the split in data. So every time you run this line of code, the same data will be set to `x_train, x_test, y_train, y_test` instead of random numbers.

```
x_train, x_test, y_train, y_test = train_test_split(x, y,
train_size = .8, random_state = 42)
```

Now, we will use the parameters found using `GridSearchCV` as our hyperparameters to train our model.

```
model = RandomForestClassifier(criterion="gini", max_
depth=5, n_estimators = 200,  random_state=42)
model.fit(x_train,y_train)
```

We will use our new model to predict the quality of wine using the features from our testing dataset.

```
predictions = model.predict(x_test)
```

```
accuracy_score(y_test, predictions)
```

The accuracy score is `0.56875`.

We can also see which features are really important to the model when making its predictions, which can be seen in Figure 6-9.

```
feature_importances = model.feature_importances_
feature_names = wine_data.columns
sorted_importance = feature_importances.argsort()
plt.barh(range(len(feature_importances)), feature_
importances[sorted_importance])
```

```
plt.yticks(range(len(feature_importances)), [feature_
names[i] for i in sorted_importance])
plt.ylabel("Features")
plt.xlabel("Importance")
plt.title("Title: Model Feature Importance")
```

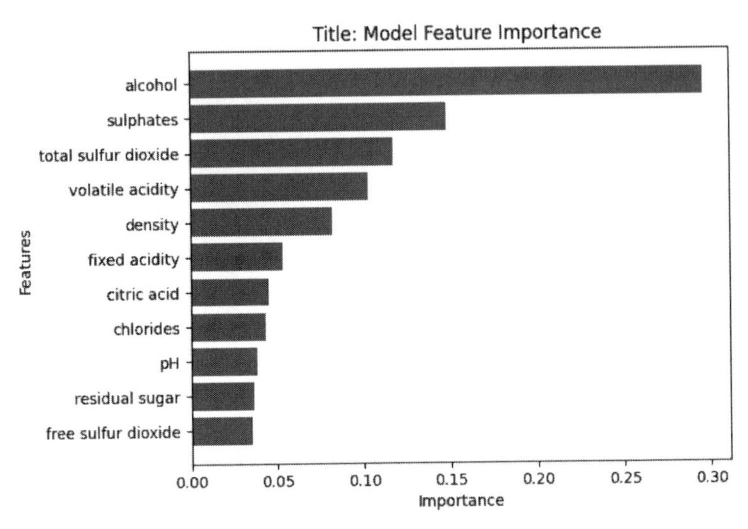

Figure 6-9. *A chart showing the features the model found most important when making its predictions*

Monitoring Your Model

Machine learning pipelines automate the process, from data preprocessing to model training, evaluation, and deployment. This ensures that each step is carried out consistently and efficiently, which is vital for reliable model evaluation.

Pipelines streamline the evaluation process by keeping track of metrics used to evaluate your model. This makes it easier to track and compare the performance of different models you deploy over time. By systematically applying the same evaluation criteria to each model, pipelines help identify the best performing model more reliably.

MLflow is an example of a platform that integrates these principles, among others, into a comprehensive solution for managing machine learning projects.

Examples Where Machine Learning Models Can Be Useful

1. Credit Card Fraud Detection
   In a credit card fraud detection scenario, evaluating the model's performance is critical due to the imbalance in the dataset (few fraudulent transactions compared to legitimate ones). Techniques like precision, recall, F1 score, and ROC-AUC can be used to measure the model's effectiveness in identifying fraud.
2. Predicting Housing Prices
   In regression tasks like predicting housing prices, metrics such as MAE, MSE, and R squared can be used to evaluate model performance. These metrics can help in understanding the accuracy of predictions.

Conclusion

Evaluating machine learning models is crucial to ensure their reliability, accuracy, and generalizability to new data. We can get a comprehensive understanding of a model's performance by using various tools and techniques such as train-test-split, cross-validation, and a range of evaluation metrics. Through examples and use cases, we've seen how these methods can be applied to different types of problems. Proper evaluation is not just a best practice but a necessity for deploying robust machine learning solutions.

CHAPTER 7

When to Use Machine Learning Models

Machine learning models have become a powerful tool for solving complex problems in various fields. These days, it seems that everyone is looking for ways to add machine learning either for personal reasons or in an attempt to advance their business. However, applying machine learning models to every problem is unnecessary. While machine learning can offer powerful solutions, it's important to recognize not all challenges require such advanced techniques. Sometimes, simpler, rule-based systems or other traditional methods might be more efficient and effective. Overcomplicating a solution with machine learning can lead to unnecessary complexity without delivering the desired value.

Imagine you have a problem to solve – perhaps optimizing an existing system, launching a new business, or enhancing customer experience. You might consider a machine learning model, but assessing whether this approach is the best fit is essential.

Introduction to When to Use Machine Learning Models

Machine learning has revolutionized various fields by providing systems the ability to learn and improve from experience without being explicitly programmed. From healthcare and finance to retail and entertainment,

M. Attobrah, *Essential Data Analytics, Data Science, and AI*,
https://doi.org/10.1007/979-8-8688-1070-1_7

machine learning is integral to driving innovation and efficiency. In Chapter 5, we discussed different types of machine learning models. However, the decision to employ machine learning should be made judiciously, as it involves complex algorithms, substantial data, and computational resources. This chapter delves into the criteria and contexts in which machine learning models are beneficial, ensuring that their application yields optimal results. This chapter explores when a machine learning model might be the right choice and provides guidance and alternative choices that may better serve your needs.

Importance of Understanding When to Use Machine Learning Models

Understanding when to deploy machine learning models is crucial for several reasons:

1. Resource Optimization

   Machine learning projects require significant investment in data collection, processing, and algorithm research/development. Correctly identifying the need for machine learning ensures that resources are utilized effectively.

2. Problem Suitability

   Not all problems are suitable for machine learning solutions. Some might be better addressed through simpler statistical methods or rule-based systems. Misapplication can lead to inefficiencies, poor outcomes, and unnecessary expenses.

3. Performance and Scalability

   Machine learning models can provide scalable solutions to problems involving large datasets and complex problems by leveraging their ability to do things like process vast amounts of data and automate decision-making. Knowing when to use machine learning can help ensure these strengths are utilized for improved performance. As you scale, monitoring model performance becomes critical. In Chapter 6, we discussed various types of model performance metrics. These metrics help provide a clear understanding of how well the model is performing, helping to identify issues like model drift or model performance degradation. These metrics also help to compare different models to help pick the right one for your business needs.

4. Innovation and Competitive Edge

   Proper use of machine learning can lead to innovative solutions, providing a competitive edge in the market. It can automate processes, uncover insights, and enhance decision-making capabilities.

Assessing the Necessity and Value of Machine Learning Models

Identifying the Right Problems

Sometimes, it's best to solve problems in the simplest way possible. When deciding whether to use a machine learning model, it's essential to consider the complexity and frequency of updates the model will require. A machine learning model may not be necessary if a problem can be solved with a straightforward one-and-done solution, such as simple calculations. Machine learning models are best suited for situations where continuous learning and frequent updates are needed, which is not the case for every problem.

Machine learning models shine in scenarios that demand constant evolution and adaptation. These include significant, complex problems that require ongoing adjustments, open-ended challenges that grow over time, situations where the correct solution changes frequently, and problems that are inherently difficult and push the limits of our capabilities as humans. If your problem falls into one of these categories, a machine learning model may be the right choice.

Navigating Open-Ended Challenges

Some challenges are enormous and open-ended, meaning they don't have a single, fixed solution. These problems continue to evolve, requiring ongoing work without a clear end. For example, curating recommendations for streaming services involves dealing with new information daily.

In such cases, a one-time solution that you set up and leave alone is unlikely to be effective. Instead, these scenarios call for machine learning models that can adapt and grow over time, continuously learning and updating as new data emerges. If your problem is well defined and finite, a machine learning model may not be necessary. However, for immense, ongoing challenges, machine learning models could be the right choice to keep up with the ever-changing landscape.

An example of a machine learning model from the list of types of models we discussed in Chapter 5 that is well suited for open-ended challenges is reinforcement learning models. These models learn by interacting with their environment, making them ideal for tasks that require ongoing improvements, such as optimizing recommendations for e-commerce platforms or streaming services and managing autonomous vehicles.

Adapting to Change in Dynamic Environments

Problems that involve changes can be tricky because what works well today might not be effective tomorrow. When dealing with issues where conditions change frequently or unpredictably, a machine learning model can be very useful. These models can learn and adjust quickly to new situations, ensuring they remain effective as things evolve. If your field changes slowly or in predictable patterns, a simpler solution may suffice. However, if you face rapid or unexpected changes, a machine learning model could be the best way to keep up and stay relevant.

Solving Tough Problems

Some problems are so complex that even humans struggle to solve them perfectly. Examples include excelling in complex games and understanding foreign languages. Machine learning models have made significant strides to address these challenging problems, but it has taken years of development and data collection. Machine learning models can be advantageous in specific tasks where vast amounts of data need to be processed or where patterns can be extracted from complex information. By continuously integrating user feedback and refining the model, these advanced solutions can make real progress in tackling more inherently challenging issues.

However, there are still areas where human intuition and expertise are key, for example, tasks that require creativity, emotional intelligence, or deep contextual understanding.

Critical Conditions for Successful Machine Learning Models

It's not just about having challenging problems for machine learning to be effective. The problem also needs to meet a few critical criteria:

- The model must be able to gather and learn from data generated from using it over time.
- The model should aim to achieve a significant goal.
- The overall effort should be worth the investment.

Critical Conditions for Successful AI Products

Machine learning models are rarely perfect and make mistakes. What's crucial is that they provide enough value to outweigh their errors. For a machine learning model to be viable, it should offer significant benefits that surpass the costs of its mistakes. For example, a model that optimizes grocery shopping for customers might occasionally make small errors that cause slight delays. As long as these errors are infrequent and the system overall saves users time, it can be considered valuable.

In contrast, a model used for critical tasks, like controlling surgical robots or helicopters, must be much more reliable since mistakes can have severe consequences. Even in such high-stakes situations, the model doesn't need to be flawless but should be significantly better than alternatives. The key is that a machine learning model must offer enough positive outcomes to justify its use, even if it's not perfect.

For a machine learning model to be seen as most effective, it must be able to directly influence the desired outcome. This means it should either automate processes or provide information and options that help users achieve their goals. The model works best when it can make actions that have a clear and immediate impact on the product's goal. If the model's actions lead to positive results, it should be evident that the model performed well, and if the outcomes are negative, it should be clear that the model's performance was lacking. The fewer external factors that affect the results, the more effective the machine learning model can be.

When deploying machine learning models, it's important to find the right balance – too broad a goal can make it difficult to connect a model's actions to its outcomes. In large, complex products, multiple machine learning models can be used for different tasks. You could use one model for multiple tasks, but the feedback from users would not relate directly enough to the specific decisions the machine learning model needs to make to get better.

Enhancing Your Model by Leveraging the Data from Interactions with It

Machine learning models improve over time by learning from how end users interact with them. When users engage with your product that incorporates a machine learning model, it should collect data about their actions, the results they experience, and their feedback. This is part of telemetry data, which is discussed in Chapter 9. This data is crucial because it helps the machine learning model become smarter and more effective, making it more valuable for the end users.

For example, to enhance the machine learning model, it's essential to record detailed information about user interactions. This involves tracking things like

- What users see
- How they respond
- And whether their experience is positive or negative

By analyzing this data, machine learning models can be continuously refined to better meet the end users' needs. This will enhance your reputation and financial prospects overall for you and your company.

Cost Considerations for AI Products

The costs associated with AI products differ from those of traditional methods. There are three main components of these costs:

- Machine Learning Model: The component of the product that makes decisions and determines the best actions
- Implementation: Covers the actual development and integration of the backend system, model, and any other components, including the user interfaces
- Orchestration: Involves managing the product throughout its life cycle to ensure it continues to meet goals and handle errors

In general, the machine learning model component of the AI product is often less expensive because these models can automatically generate the output of the product instead of a human and it can produce the necessary training data through user interactions. This reduces the need for manual data collection and can overall save money. However, implementation costs for machine learning models can be comparable to or slightly higher than traditional methods due to the added complexity of creating, integrating, and organizing these components. The orchestration costs also tend to be higher, as they involve ongoing management tasks like tuning, updating, and correcting errors.

When to Consider Alternatives

Sometimes, you might have a situation involving users, data, and decision-making, but you're unsure if a machine learning model is necessary. This uncertainty can stem from several reasons: you may doubt whether your problem is complex enough to require a machine learning model, you may question whether the effort and costs of implementing such a model are justified, or you prefer to address issues incrementally rather than committing to researching and developing machine learning models from the start.

It's perfectly fine to explore other methods for solving challenging problems, and machine learning models aren't the only solutions.

Tools and Techniques of When to Use Machine Learning Models'

Is your AI project viable and worth pursuing?

Evaluate your AIs project feasibility with this table

Business Viability	Data Viability	Implementation Viability
Is there a clear problem?	Is thee data quality sufficient or do you at least have the team in place to get that done?	Do you or your team have the required technology and skills?
Is the company willing to invest and change?	Is there a sufficient quantity to train or finetune the model?	Does it make sense to use the model where and how you plan to use it?
Is there a sufficient ROI or impact?	Can you get access to the required data that measures what you care about?	Can you execute the model as required in a timely and efficient manner?

Figure 7-1. *Sample checklist of evaluating an AI project's viability*

More Key Indicators You Can Use for Deciding on Employing Machine Learning Models in Your Product or Systems

1. Large and Complex Datasets

   When the data size is substantial and contains intricate patterns that are not easily captured by traditional methods, machine learning is a valuable tool.

2. Predictive Analysis

   If the goal is to predict future trends based on historical data.

3. Patten Recognition

   Tasks involving pattern recognition, such as image recognition, anomaly detection, and clustering, can be prime candidates for machine learning.

4. Automation of Repetitive Tasks

   When repetitive tasks need automation, such as in manufacturing or data extraction, machine learning can provide efficient solutions.

5. Adaptation and Learning

   If the system needs to adapt and improve overtime based on new data, a machine learning model's ability to learn from data makes it suitable.

Techniques and Tools

In Chapter 5, we learned about different machine learning algorithms and techniques. Below are some things to consider when deciding which algorithm to use.

1. Supervised Learning

   Used when the outcome variable is known. Techniques include linear regression, logistic regression, decision trees, and support vector machines.

2. Unsupervised Learning

   Applied when the outcome variable is not known. Techniques include clustering (k-means, hierarchical) and association rules.

3. Reinforcement Learning

   It is used in dynamic environments where the model learns from actions and their outcomes. Applications include automotive vehicles and game play.

4. Deep Learning

   A subset of machine learning focusing on neural networks with many layers. It is particularly useful for tasks involving image, speech, and text processing.

5. Natural Language Processing (NLP)

   Techniques for understanding and generating human language are crucial for applications like chatbots, sentiment analysis, and language translation.

EXAMPLE WITH CODE

Here is an example where machine learning would be unnecessary.

A company wants to assign unique employee ID numbers to new hires. The ID numbers can be simple numbers that increase sequentially, starting from 10001. The numbers would increase by 1 for each new employee that joins the company.

This is unnecessary because assigning sequential ID numbers is a straightforward task. Each new employee will get the next available number in the sequence.

Here is an example using Python.

Let's say you have the ID saved in a database or Excel spreadsheet; you would begin with the last employee ID assigned, which, in this example, is 10001:

```
last_employee_id = 10001
```

Create a function to increment the ID number based on the previous one assigned:

```
def generate_employee_id(last_id):
  return last_id + 1
```

Use the function to generate the next employee ID for your new hire:

```
new_hire_id = generate_employee_id(last_employee_id)
```

Even using Python to implement this may be overdoing it, but since this is a technical book, I decided to show you a code implementation!

Here is an example where machine learning can be useful.

Let's say you have a wine store and have a dataset of customer information. You want to group customers into segments based on their purchasing behaviors to target them with personalized market strategies. The data we will be working with in this example has a little over 2000 rows.

Import required libraries.

```
import pandas as pd
from sklearn.cluster import KMeans
from sklearn.preprocessing import StandardScaler
import matplotlib.pyplot as plt
import seaborn as sns
```

Load the dataset. This dataset can be found on Kaggle: `https://www.kaggle.com/datasets/imakash3011/customer-personality-analysis/data`

```
data = pd.read_csv("marketing_campaign.csv", sep="\t")
```

This data has 29 columns that describe the customer, for example, marital status, income, number of kids, amount spent on wine, etc.

```
data.columns
```

After typing the code above, you will see a list of column names as seen below in the brackets.

```
Index(['ID', 'Year_Birth', 'Education', 'Marital_
Status', 'Income', 'Kidhome', 'Teenhome', 'Dt_
Customer', 'Recency', 'MntWines', 'MntFruits',
'MntMeatProducts', 'MntFishProducts', 'MntSweetProducts',
'MntGoldProds', 'NumDealsPurchases', 'NumWebPurchases',
'NumCatalogPurchases', 'NumStorePurchases',
'NumWebVisitsMonth', 'AcceptedCmp3', 'AcceptedCmp4',
```

```
'AcceptedCmp5', 'AcceptedCmp1', 'AcceptedCmp2', 'Complain',
'Z_CostContact', 'Z_Revenue', 'Response'],
  dtype='object')
```

In this example, we will focus on grouping the customers by their income (Income) and the amount of wine they purchased in the last two years (MntWines). Like most datasets you will encounter in the real world, this data needs to be cleaned. For this example, we will keep it simple by filling any empty values within the Income and MntWines columns with a 0.

```
features = data[["Income","MntWines"]].fillna(0)
```

As you can see in Figure 7-2, some values are significantly different from each other. Here, we will transform the values to have a common scale. This will allow the k-means algorithm to focus more on the patterns of the data as opposed to the size, especially since this is a distance-based algorithm.

	Income	MntWines
0	58138.0	635
1	46344.0	11
2	71613.0	426
3	26646.0	11
4	58293.0	173
...	...	...
2239	52869.0	84

Figure 7-2. *Sample of data within the two columns*

```
scaler = StandardScaler()
scaled_features = scaler.fit_transform(features)
```

In this example, we will group the customers in five groups.

```
kmeans = KMeans(n_clusters=5)
clusters = kmeans.fit_predict(scaled_features)
```

After identifying the clusters, we will place those values back in our table for further analysis. Here we are putting the values of the clusters in a new column called `Clusters`. The clusters range from 0 to 4.

```
data["Clusters"] = clusters
```

***Figure 7-3.** Sample of dataset after adding cluster values in the last column*

Now, let's analyze our data based on the clusters.

```
sns.barplot(data = data, x="MntWines", hue="Clusters")
```

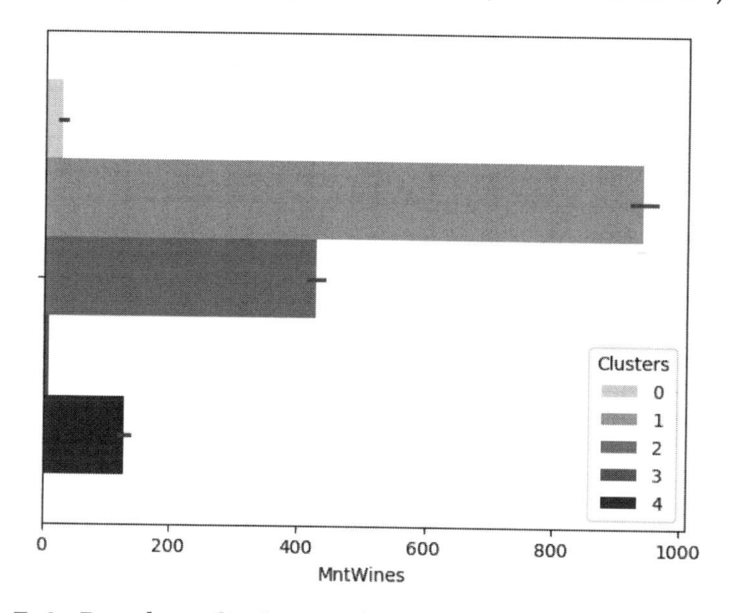

***Figure 7-4.** Bar chart displaying the number of wines purchased based on the clusters*

Here we can see the people in cluster 1 bought the most wine.

```
sns.countplot(data = data, x="Kidhome", hue="Clusters")
```

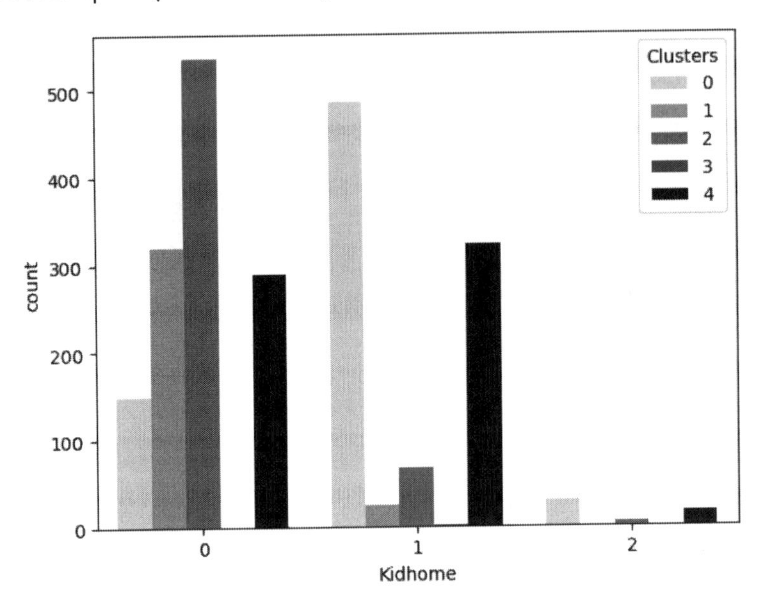

***Figure 7-5.** Bar chart displaying the number of kids each cluster has*

Further analysis shows that most people in cluster 1 do not have kids. So when creating a marketing campaign to increase wine sales from your store, you would probably want to steer away from campaigns encouraging customers to relax with a bottle of wine while their kids are at the babysitter because that will not resonate with your most paying customer group.

Examples Where Machine Learning Models Can Be Useful

Finance: Fraud Detection

In the financial sector, machine learning models can be crucial for detecting fraudulent transactions. By analyzing transaction patterns, these models can identify anomalies that suggest fraudulent activity. Financial institutions can use these insights to prevent fraud and safeguard assets.

Retail: Personalized Recommendations

Retailers can leverage machine learning models to provide personalized recommendations to customers. By analyzing purchase history, browsing behavior, and customer preferences, these models can enhance the shopping experience and increase sales.

Conclusion

Machine learning models have immense potential to transform various industries by providing advanced solutions to complex problems. However, the decision to use machine learning should be based on a thorough understanding of the problem at hand, the availability of data, and the potential benefits. By recognizing the appropriate context for machine learning development and deployment, you can harness its power to drive innovation, efficiency, and competitive advantage.

CHAPTER 8

Where Machine Learning Models Live

When creating an AI-driven product, you need to decide where the machine learning model and information needed to make predictions or decisions will come together and then how you will send those outputs back to the end user. One option is to have the machine learning model live directly on the user's device. If you do this, you will need to think about how to keep the model updated for all users. Another option is to have the model work in the cloud. In this case, you will need to figure out how to send the necessary data from the user's device to the cloud quickly and in an inexpensive manner that works smoothly.

This chapter explains some things to keep in mind when deciding where a machine learning model should live. It discusses important factors and why they matter for different types of AI-driven products.

Introduction to Where Machine Learning Models Live

Machine learning has significantly transformed various industries by providing AI-driven products to complex challenges. However, the effectiveness of these solutions is heavily influenced by the specific environment where the machine learning models are deployed and

M. Attobrah, *Essential Data Analytics, Data Science, and AI*,
https://doi.org/10.1007/979-8-8688-1070-1_8

executed. To ensure optimal performance, scalability, and reliability, it is essential to comprehend the various environments that can host these models. This chapter explores the different types of environments, including cloud services, on-premise options, and hybrid systems, each providing unique advantages and considerations for deploying machine learning models within them.

Key Considerations to Think About When Deciding Where Your Machine Learning Models Should Live

The environment where a machine learning model is deployed can significantly impact its performance and usability. When figuring out where to put machine learning models in an AI-driven product, here are some important points to keep in mind:

- Timely Updates

  How long will it take to update the machine learning model?

  - Machine learning models need to be updated regularly to ensure users benefit from the latest improvements. When these updates can be implemented directly on the device users are interacting with, they can experience the benefits much more quickly. However, if the model is hosted remotely and updates require Internet access, the process may be slower, potentially leading to significant issues if the delays result in costly errors.

For example, if you launch a product that gains popularity, it's essential to capture user feedback and interaction data to refine your model. In such cases, being able to rapidly update the model for all users is crucial to maintaining momentum and delivering a high-quality experience. This is especially true in fast-changing scenarios, such as combating cybersecurity threats, where delays in updates could render the model ineffective.

Conversely, in situations where the problem is evolving more slowly, such as when constructing a new building, delays in updates may not be as critical. A GPS can still function adequately during the construction period without immediate updates.

The urgency of updates becomes apparent when a lack of timely changes can lead to costly mistakes. If you find yourself receiving frequent customer complaints but are unable to address them promptly due to slow model updates, this could damage your reputation and financial standing and even cause significant stress.

Strategies for Updating Models

Below are a few different strategies to help ensure timely updates:

- Continuous Integration/Continuous Deployment (CI/CD) Pipelines

  This approach allows developers to automatically test, build, and deploy model updates to production environments. CI/CD pipelines ensure rapid updates and minimize errors by automating much of the update process.

- Over-the-Air (OTA) Updates for Edge Devices

  For models running on edge devices, OTA updates allow changes to be pushed directly to the device with minimal user intervention. This method ensures that even devices typically running offline can stay up to date when they periodically connect to the Internet, minimizing security risks and improving functionality without much delays.

- Cloud-Based Training

  In cloud environments, models can be retrained and redeployed quickly as new data becomes available. Cloud platforms offer scalability, allowing updates to be applied simultaneously across multiple locations or users.

- Latency

  The physical proximity of the deployment environment to the end user can affect the response time of the model. This can affect how quickly the model will generate a response when it is needed.

  - Your AI-driven product must gather the required data, perform preprocessing, and extract the necessary features to serve as input for your machine learning model. After the model generates a response, you may need to apply postprocessing before delivering the final output to the end user. Each of these steps can take time.

If all processing happens on the same device, the delay is minimal. However, the process can take longer if the machine learning model is hosted elsewhere, such as in the cloud. This latency can be particularly problematic in situations where the correct response needs to be generated rapidly, such as during an emergency. For example, consider an autonomous helicopter that suddenly encounters an obstacle. If there's a delay in processing, the helicopter may not be able to avoid the obstacle in time, leading to potentially dangerous consequences.

- Cost

  The cost of deploying and maintaining machine learning models varies with the chosen environment. Running an AI-driven product requires utilizing resources, which all cost money. Here are some key points to consider:

  - Cost of Distributing the Machine Learning Model

    - Distributing machine learning models involves sending updates to users, which costs money for both you and your customer. The cost depends on how many customers need the update, the size of the update, and how often updates are sent.

    - For users with Wi-Fi, this might not be a big deal. However, for those using mobile devices or who have limited data, the cost of downloading updates can be significant.

- Cost of Executing Machine Learning Models
  - Running the machine learning model has costs, especially if it needs to send and receive data from the cloud. The size and frequency of these data exchanges can add up.
  - Machine learning models on an end user's device use their resources, such as the CPU, RAM, and GPUs. While this can save you money as the product developer, some machine learning tasks can be too demanding for smaller devices, such as mobile phones. In these cases, having the model on the cloud might be more efficient.
- Offline Mode

  What happens if the end user is not connected to the Internet? You should also consider if the model should work even if there is no Internet connection.

  Sometimes it's okay if the model doesn't work offline. However, in some cases, like life-saving equipment, it is crucial the product works without an Internet connection. Some portion of the machine learning capabilities should live on the end user's device. It can be the full version or a back-up version to keep the product working until the Internet connection is back on.
- Scalability

  Cloud environments often offer scalable resources that can be adjusted based on demand, while on-premise solutions might require more upfront investment and maintenance.

- Security

  Different environments provide varying levels of security. For instance, sensitive data might be better protected in a private on-premise setup compared to a public cloud.

Deciding Where the Machine Learning Should Live

Built-In Machine Learning Model

You can include a machine learning model directly in your product without any extra setup. You can collect training data, create the model, add it to your software, and then release it.

Advantages

- It can be cheaper and simpler to set up.
- It can be good enough for many problems that do not change a lot and are not very complex.
- It can be updated through traditional software updating methods.
- You could get feedback through reviews and other methods.

Disadvantage

- Updating the model is more challenging.

Cloud-Based Machine Learning Model

Cloud-based machine learning means the model runs in the cloud, not on the user's device. When a user needs something, the application sends the information to the cloud, which processes it and returns the results.

Advantages

- The model can be updated easily and quickly.
- It is easier to track and log data.

Disadvantage

- As more people use it, the cloud will need to handle many requests quickly. If many people are using it at the same time, the system has to be powerful enough to keep up.

Stored Machine Learning Results

Stored machine learning results involve figuring things out ahead of time and saving those results for later.

Example: Imagine a book recommendations service that figures out which books in a finite list or slow-updating list are best for different genres and readers. It does this detailed analysis and stores the results. When a new reader looks for book suggestions, the application quickly uses the stored recommendations without having to redo all the analysis.

Advantages

- It can run complex models that would be too slow if done in real time.
- It can quickly provide results since they are already prepared and stored.

Disadvantages

- It works best if the results can be looked up based on clear information, like specific location or product.
- It is not effective if the correct answer can change quickly.

Mixing Different Approaches

Sometimes, combining different methods to get the best results is helpful. Here are some examples:

- One machine learning model can be used to do a deep analysis on a specific task, while another model on the device can handle other things.
- An application can use a machine learning model on the device most of the time but double-check with a more powerful model in the cloud when making very important decisions.

Combining methods can help balance out their individual weaknesses. However, it can also make the application more complicated. If there is a mistake, it can be tricky to know which part caused it. It could be the model on the device, the model in the cloud, or a combination of things.

Tools and Techniques

Cloud Services

Cloud platforms like Amazon Web Services (AWS), Google Cloud Platform (GCP), and Microsoft Azure provide robust environments for deploying machine learning models. These platforms offer various services, such as

- Managed Machine Learning Services: AWS SageMaker, AWS Bedrock, Google AI Platform, and Azure Machine Learning provide end-to-end solutions for building, training, and deploying models.
- Serverless Computing: AWS Lambda and Azure Functions allow for running code in response to events without managing servers.
- Containerization: Tools like Docker and Kubernetes enable scalable and flexible deployment of machine learning models in containerized environments.

On-Premise Options

For organizations with specific requirements, on-premise deployment offers control over the hardware and software environment. Key tools and techniques include

- Virtual Machines: Creating isolated environments using VMware, for example
- Dedicated Hardware: Using specialized hardware
- Custom Infrastructure: Building and maintaining a tailored infrastructure to meet specific performance and security needs

Edge Devices

Deploying machine learning models on edge devices like smartphones and IoT devices can minimize latency and reduce data transfer costs. Techniques for edge deployment include

- Model Compression: Techniques like quantization and pruning to reduce the model size and computation requirements.
- Frameworks for Edge Deployment: TensorFlow Lite and PyTorch Mobile are designed for running models.

Hybrid Systems

Combining different environments to leverage the strengths of each can be a strategic approach. For example, initial model training could occur in the cloud, while inference might take place on edge devices. Key considerations for hybrid systems include

- Data Synchronization: Ensuring that data is consistent across environments
- Network Architecture: Designing robust and secure communication channels between different deployment environments

Examples of Machine Learning Deployment Strategies for Industry Needs

Retail: Cloud-Based Recommendation Systems

Retailers can leverage cloud services to deploy recommendation systems that analyze customer data and provide personalized shopping experiences.

Updates and Monitoring: In a cloud-based system, models can be continuously retrained and redeployed without impacting the customer experience. Cloud environments also allow retailers to monitor how well the recommendations system performs across regions, adjusting models as needed for varying preferences or market conditions.

Effectiveness: Cloud deployment enables scalability, allowing the system to handle large customer bases and real-time data analysis. However, latency can impact effectiveness during peak shopping periods if the cloud infrastructure isn't optimized to handle surges.

Healthcare: Edge AI for Real-Time Diagnostics

In the healthcare sector, deploying machine learning models on edge devices such as portable diagnostic equipment can provide real-time analysis and reduce the need for Internet connectivity, for instance, a company developing a portable ultrasound device with embedded AI capabilities to assist in diagnosing conditions on the spot.

Manufacturing: Hybrid Systems for Predictive Maintenance

Manufacturers can use hybrid systems where data from sensors on equipment is processed locally to detect anomalies in real time, while more extensive analysis and model retraining can occur in the cloud. This approach would balance the need for low-latency monitoring with the computational power of cloud resources.

Updates and Monitoring: On-site models can be updated through OTA, while the cloud component ensures the models are retrained with the latest data from all the equipment. Monitoring these systems involves tracking both the performance of the local anomaly detection models and the more complex cloud models used for deeper analysis.

Effectiveness: A hybrid approach balances the need for low-latency anomaly detection with the computational power of cloud resources. The local model ensures real-time monitoring, while cloud-based retraining keeps the overall system adaptive to changing conditions. However, balancing synchronization between edge and cloud systems is crucial to ensure smooth operations without data bottlenecks.

Conclusion

The deployment environment for machine learning models is a critical factor influencing their effectiveness, performance, and cost-efficiency. Understanding the strengths and limitations of various environments - cloud, on-premise, and hybrid systems - allows organizations to make informed decisions tailored to their specific needs. By leveraging appropriate tools and techniques, businesses can optimize the deployment of their machine learning models to achieve the best possible outcomes.

CHAPTER 9

Telemetry

Introduction to Telemetry

Telemetry, derived from the Greek roots "tele," meaning remote, and "metron," meaning measure, is the process of recording and transmitting the readings of instruments and devices to a remote location. In the context of machine learning, telemetry plays a crucial role in monitoring, diagnosing, and optimizing the performance of machine learning models. By leveraging telemetry, data scientists and machine learning engineers can gather insights into model behavior, system performance, and user interactions, leading to more robust and efficient systems.

Telemetry in machine learning involves collecting data from various sources, such as model performance metrics, system logs, and user feedback. This data can then be transmitted to a central system for analysis and visualization. The insights gained from telemetry can be used to fine-tune models, detect anomalies, and ensure the overall health of the machine learning pipeline.

M. Attobrah, *Essential Data Analytics, Data Science, and AI*,
https://doi.org/10.1007/979-8-8688-1070-1_9

Importance of Telemetry

The importance of telemetry in machine learning cannot be overstated. It provides a continuous feedback loop essential for maintaining machine learning models' reliability and accuracy. Here are some key reasons why telemetry is vital:

1. Model Performance Monitoring

   Telemetry allows for real-time monitoring of model performance, enabling the detection of issues such as data drift, model decay, and performance degradation. This ensures that models remain accurate and reliable over time.

2. Anomaly Detection

   Analyzing telemetry data allows for the detection of anomalies in model behavior or system performance. This can help identify and address potential issues before they impact the end users.

3. System Optimization

   Telemetry provides insights into system performance, including resource utilization, latency, and throughput. This information can be used to optimize system configurations and improve inefficiency.

4. User Feedback Integration

   Telemetry can capture user interactions and feedback, which can be invaluable for improving model accuracy and user experience. Understanding how users interact with the system allows for better model tuning and personalization.

Telemetry Tools and Techniques

Various tools and techniques can be implemented to capture telemetry data from AI systems. These tools can be broadly categorized into data collection, transmission, and analysis.

Data Collection Tools

1. Loggers, such as Python's "logging" module, record events and metrics from machine learning models and AI systems. These logs can capture detailed information about model predictions, errors, and performance metrics.

2. Prometheus and OpenTelemetry provide instrumentation capabilities to collect metrics from AI systems. These tools can be integrated into machine learning pipelines to collect data on model and system performance metrics.

Data Transmission Tools

1. Message brokers like Kafka and RabbitMQ transmit telemetry data from various sources to a central system. They provide reliable and scalable data transmission solutions.

2. REST APIs can transmit telemetry data to centralized systems. They offer flexibility and ease of integration with various data sources and destinations.

Data Analysis Tools

1. Dashboards and Visualization Tools

   Tools like "Grafana" and "Kibana" allow for the visualization of telemetry data through interactive dashboards. They provide powerful features for creating graphs, charts, and alerts based on telemetry data.

2. Data Processing Frameworks

   Frameworks like Apache Spark and Flink can process and analyze large volumes of telemetry data. They offer real-time and batch processing capabilities, making them suitable for various use cases.

   Application Monitoring Tools

   Tools like "New Relic" and "Datadog" offer comprehensive solutions for monitoring the performance of applications, including machine learning models. They provide features of collecting metrics, tracing requests, and visualizing data.

Model Decay, a.k.a. Model Drift

Model drift occurs when machine learning models lose their ability to perform well due to shifts in the data they're trained on. This can result in incorrect predictions and bad decisions. Since the world is constantly changing, models need to be regularly reviewed and updated.

An example of this change in data is data drift. In machine learning, this happens when the data a model was trained on starts to change over time, making the model less accurate. If the machine learning model is not ready for these changes, it might not work as well as before.

For example, in e-commerce and language translation, model drift can be a challenge. An e-commerce recommendation system may start giving irrelevant product suggestions if customer preferences or trends change over time. Similarly, a change in language usage, slang, or cultural references can impact the performance of a model. In both cases, performance would degrade without regular updates to the model and monitoring.

There are tools and frameworks specifically designed to detect and handle model drift. For example, Python packages such as scikit-multiflow and River support drift detection by monitoring the data and alerting when significant shifts occur. These tools help to identify when it is time to retrain models and adapt to new patterns in the data.

Imagine you taught a robot to recognize fruits using apples and oranges. If you suddenly show it bananas, the model might get confused because it wasn't trained on that new fruit. This could happen because the way data is collected changes or because the group of people or things being studied is different. If a model was trained using data from one country but is later used in another country where things are different, it might not give good results.

Types of Drift

Imagine you are teaching a machine learning model to predict if it will rain tomorrow.

- Target Drift: This is like changing the rules of the game after the model has learned to play. If you suddenly decide that "drizzle" also counts as rain, the model might get confused because it only expected "rain" before.

- Concept Drift: This is like the world changing around the model. Maybe it learned about rain in warm weather, but then it has to predict rain in a cold place like with snow. That's a big change!
- Data Drift, a.k.a. Covariate Shift: This is like changing the information you give the model. If you used to tell the model the temperature and humidity to predict rain, but now you give it temperature, the model might not be as good at predicting.
- Label Drift: This is like changing the answers key after the test. If you said "yes" to rain when it drizzled before, but now you say "no," the machine will get confused about what counts as rain.

Ways to Detect Drift

To deal with drift, we need to design machine learning models that can notice when things are changing and adjust to those changes. This might mean regularly checking how well the model is working and retraining it with new data or even creating models that can quickly adapt when they notice something is different.

So let's continue to focus on the idea that the model is made for weather. Over time, the weather patterns can change, and if your model doesn't adapt, its predictions will become less accurate.

- Regular Check-Ups: We need to regularly test how well our AI model is performing. This helps us spot any problems early on.

- Detecting Changes: Special tools can help us identify when the data the model is using starts to change. This is like noticing a sudden shift in weather patterns. There are algorithms designed specifically to detect drifts.
- Preventing Problems: We can prepare our AI model for changes by showing it a wide variety of data. We could use various techniques for this. For example, data augmentation and synthetic data generation can assist in exposing the model to a wider range of representative data. This is similar to teaching a child about different types of weather.
- Updating the Model: If your model starts making mistakes, we can teach it new information to improve its accuracy by fine-tuning it. This is like giving a child extra lessons when they struggle with a subject.

Examples of How Telemetry Can Be Used in the World

E-commerce Recommendation Systems

In an e-commerce platform, telemetry can be used to monitor the performance of a recommendation system. The system collects data on user interactions, model predictions, and system performance. By analyzing this telemetry data, the platform can detect changes in user behavior, identify performance bottlenecks, and optimize the recommendation algorithm.

Specific telemetry data that can be collected are

- User Interactions: User clicks, purchases, product views, time spent on pages, user feedback on recommendations
- Model Predictions: Recommendations made
- System Performances: Model response time

By analyzing this data, you can

- Retrain the model based on customer behavior
- Identify biases and take corrective action
- Improve response time to enhance the customer's experience

Key Takeaways

- Improved model accuracy through continuous feedback
- Early detection of anomalies leading to timely intervention
- Enhanced user experience through better personalized recommendations

Challenges: Handling large volumes of user interaction data can be computationally intensive, requiring robust data storage and processing infrastructure. Additionally, ensuring user data privacy while collecting telemetry is critical.

Autonomous Vehicles

Telemetry is critical in the development and operation of autonomous vehicles. Vehicle sensors collect data on various parameters, such as speed, acceleration, nearby vehicles, and moving objects like pedestrians. This data is used for real-time analysis and decision-making. Telemetry helps ensure the safety and reliability of autonomous vehicles by providing insights into system performance and detecting potential issues.

Specific telemetry data that can be collected are

- Sensor Data: Speed, acceleration, steering angles
- Model Predictions: Objects detected, route optimization based on traffic conditions or objects detected
- Vehicle State: Information from cameras
- Environmental Data: Weather conditions

By analyzing this data

- Autonomous vehicles can use this data for real-time decision-making.
- You can retrain the machine learning models within the vehicle based on experiences on the road to improve their performance.
- You can identify anomalies in the data, such as sensor failures.

Key Takeaways

- Enhanced safety through real-time monitoring and analysis
- Optimization of vehicle performance based on telemetry data
- Early detection and resolution of system anomalies

Challenges: Managing the high volume of data from multiple sensors in real time is computationally demanding. Additionally, ensuring the security and privacy of telemetry data, especially when transmitted over networks, is essential to prevent cyber threats.

Healthcare Monitoring System

In healthcare, telemetry can be used to monitor patients remotely. Wearable devices collect data on health metrics like vital signs and activity levels. This data can be transmitted to healthcare providers for analysis and intervention. Telemetry enables continuous monitoring of patients, leading to timely medical interventions and improved patient outcomes.

Specific telemetry data that can be collected are

- Vital Signs: Heart rate, blood pressure
- Activity Levels: Steps taken, tremors
- Model Predictions: Disease predictions, medication effectiveness
- User Reporting: User-reported symptoms, medications taken

By analyzing this data, you can

- Retrain the machine learning models to predict the likelihood of future health events
- Identify biases and take corrective action

Key Takeaways

- Models can adapt to changes in patient conditions by analyzing new telemetry data.
- Continuous patient monitoring for early detection of health issues.

- Improved patient outcomes through timely medical interventions.
- Enhanced healthcare provider efficiency through remote monitoring.

Challenges: Managing the sensitive nature of patient data while ensuring compliance with privacy regulations like HIPAA. The sheer volume of continuous health data can also create storage challenges, especially in large-scale healthcare systems. Models can be complex and seem like a black box. Ensuring transparency in how models arrive at their predictions is important for building trust in healthcare applications.

Conclusion

Telemetry is an essential component of AI systems. It provides a continuous feedback loop that enables real-time monitoring, anomaly detection, and system optimization. By leveraging telemetry, you can ensure the reliability, accuracy, and efficiency of your machine learning models and your AI system overall.

The implementation of telemetry requires a combination of data collection, transmission, and analysis tools. With the right tools and techniques, you can gain valuable insights into model behavior, system performance, and user interactions. These insights can be used to fine-tune models and improve user experience, which enhances reputation and revenue.

As machine learning continues to evolve, the role of telemetry will become increasingly important. It will enable you to build more robust, efficient, and user-centric systems, paving the way for the future of intelligent technologies.

CHAPTER 10

Adversaries and Abuse

Machine learning has become a pivotal component in various industries, enabling advancements in automation, data analysis, and decision-making processes. However, the rise of machine learning also brings about new challenges, particularly in security. Adversaries can exploit machine learning systems, leading to significant abuse and compromising the integrity of these systems. This chapter delves into the nature of adversaries and abuse in machine learning, examining their importance, tools, techniques, and real-world cases.

Introduction to Adversaries and Abuse

Adversaries in machine learning refer to entities, whether individuals, groups, or organizations, that intentionally manipulate or exploit machine learning models to achieve malicious objectives. These adversaries aim to disrupt or deceive models for personal, financial, or strategic gain. For example, an adversary could manipulate an email spam filter by slightly altering phishing emails to avoid detection, allowing them to bypass security systems.

M. Attobrah, *Essential Data Analytics, Data Science, and AI*,
https://doi.org/10.1007/979-8-8688-1070-1_10

Abuse encompasses various activities in which adversaries compromise, deceive, or manipulate systems. Abuse also encompasses various attacks or unethical uses of machine learning systems, resulting in unintended consequences. For example, adversaries might abuse a recommendation system by using bots to artificially boost certain content, deceiving the system to prioritize irrelevant or harmful materials.

Types of Adversaries

- Data Poisoning Attacks: Intentionally putting misleading or incorrect data into training datasets, causing the model to learn incorrect patterns. This type of attack happens in the training phase. This can be done in several ways:
    - Deleting portions of the data
    - Modifying the existing data to the wrong answers
    - Adding fake data

  Example: An attacker might alter training data for a facial recognition system by injecting false images, leading the model to misidentify people.

- Evasion Attacks: Modify input data to evade detection by models. This happens in the testing/production phase by adding small changes to the data. Figures 10-1 to 10-4 show many examples of this.

  Example: An attacker could slightly alter an image of a stop sign, as seen in Figures 10-3 and 10-4, so that an autonomous vehicle's recognition system fails to identify it correctly, posing serious safety risks.

- Model Inversion Attacks: Reconstruct data that was used to train it using model outputs.

  Example: A model trained on sensitive health data could be exploited by adversaries to reverse-engineer personal medical information from the model's predictions, compromising patient privacy.

The motivations behind these attacks are diverse. They range from financial gain to causing harm. Adversaries can be malicious actors seeking to exploit vulnerabilities, like an angry ex, or competitors aiming to harm a rival's system.

Importance of Understanding Adversaries and Abuse in Machine Learning

Impact on Security and Privacy

Some machine learning models handle sensitive data, making them prime targets for adversaries. Understanding the potential for abuse is crucial for safeguarding user privacy and maintaining data integrity. To mitigate these risks, various safeguards can be implemented:

- Encryption

  Encryption helps protect data at rest and in transit. Data can be encrypted before it's used for training, ensuring that even if the data is intercepted or accessed by unauthorized individuals, it remains unreadable. For example, financial institutions can encrypt transaction data before feeding it into machine learning models to prevent unauthorized access to sensitive information.

- Access Control

  Implementing strict access control ensures that only authorized personnel have access to sensitive data and the machine learning models. Role-based access controls can limit who can view, modify, or train models with sensitive information. For example, healthcare organizations can enforce access controls so that only approved medical staff have access to patient records used in predictive health models.

- Anonymization Techniques

  Anonymization involves removing or masking personally identifiable information (PII) from datasets before they are used in machine learning models. Techniques like data masking, differential privacy, and k-anonymity can help protect user privacy by making it difficult to trace back to an individual. For example, anonymizing health records by removing or masking patient identifiers allows researchers to use the data for analysis while maintaining privacy.

Economic and Operational Implications

Exploiting machine learning systems can lead to significant financial losses and operational disruptions. For example, adversarial attacks on financial models can lead to incorrect predictions, resulting in substantial economic consequences.

Trust in Machine Learning Systems

Maintaining trust in machine learning systems is essential for their widespread adoption. Addressing adversaries and abuse is key to ensuring these systems are reliable and secure.

Ways Adversarial Attacks Can Be Used

Data Poisoning in Autonomous Vehicles

Autonomous vehicles rely heavily on machine learning models for navigation. Data poisoning could involve manipulating sensor data to cause incorrect decision-making.

Model Inversion in Healthcare

Machine learning models in healthcare can be targeted for model inversion attacks to extract sensitive patient information.

Evasion Attacks Using Adversarial Images

Let's look at some practical examples. Adversarial images are images specifically designed to trick neural networks into making incorrect judgments about the content they are seeing.

Convolutional neural networks and vision transformers are commonly used to recognize objects in images. We, as humans, recognize objects by looking at abstract features. For example, for a panda, we would look for features like black eyes and fur. Machine learning models recognize these objects differently, which makes them susceptible to adversarial images. Below are various examples of these types of images.

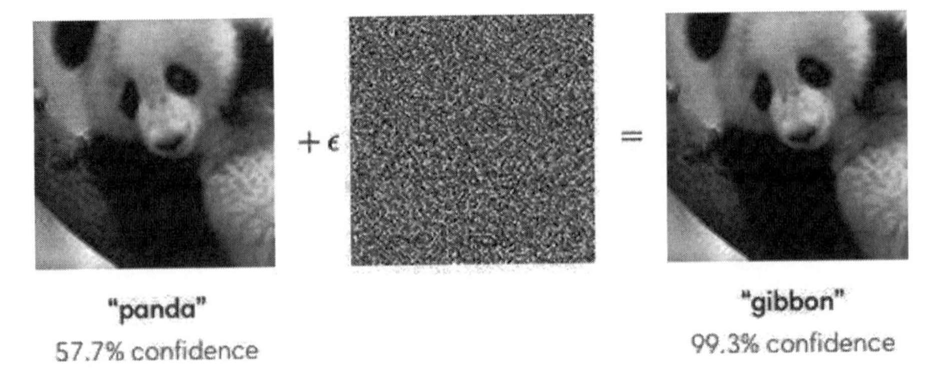

***Figure 10-1.** An example of an adversarial input tricking AI to misidentify a panda as gibbon. This image has been slightly changed by adding a noise*

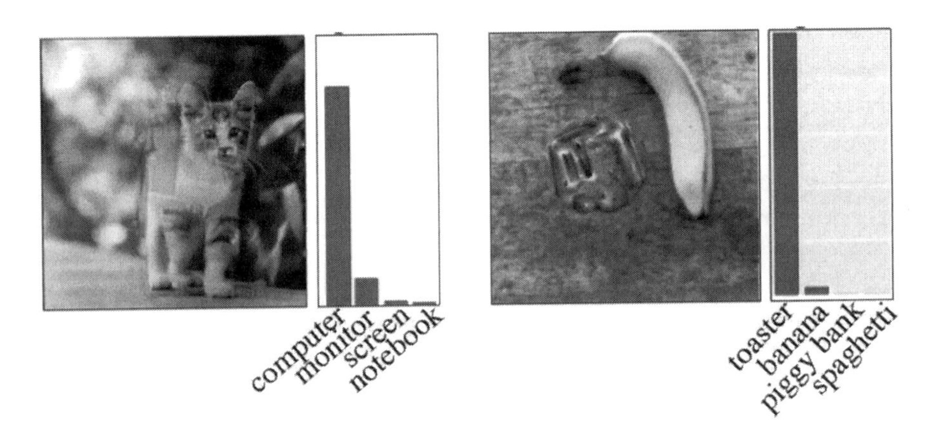

***Figure 10-2.** The image on the left is a cat but has been changed to include some boxiness. This causes the model used to identify this as a computer. The image on the right is clearly a banana, but the image has been changed to include an object on the left of it, which causes the model to identify this image as a toaster*

Figure 10-3. *In this example, art stickers and graffiti with camouflage caused a machine learning model to classify stop signs as a speed limit of 45 sign and yield sign*

Figure 10-4. *In this example, subtle changes cause the machine learning model to classify stop signs as a speed limit of 45 and right turn signs as stop signs*

Mitigating Risk

Using a machine learning model to find abusive activity could be seen as a good idea. However, it may be cheaper and faster for an attacker to change their approach to harm a machine learning model than for you to update your training dataset and model to fight against this new attack.

Here is a list of recommendations and precautions to mitigate the risk associated with and promote the responsible development and use of machine learning models.

- Training and Testing: Integrate strong adversarial training techniques during the model development phase. This process involves exposing the model to various adversarial examples within the training dataset to strengthen its defenses against evasion attacks. Rigorous testing should also be used to assess the model's robustness before deployment. This can ensure the model can effectively withstand those potential attacks. Red team testing, which will be mentioned in the "Ethical Considerations" section, is an interactive way of testing machine learning models to simulate the tactics an attacker would use to expose vulnerabilities like leaking sensitive data. The goal of this is to enhance the model's robustness.
- Data Sanitization and Validation: During the training phase, it is crucial to sanitize and validate the data to reduce the risks of poisoning attacks. This includes implementing rigorous data validation rules and mechanisms to detect and eliminate inappropriate data. Regular audits of training datasets can help identify and remove any compromised data that could undermine the model's integrity.

- Adaptive Security Measures: Employ adaptive security measures that continuously monitor and analyze model behavior during model deployment to detect unusual activities indicative of adversarial attacks. Use anomaly detection algorithms and intrusion detection systems to identify and respond to threats in real time.
- Collaboration and Information Sharing: Educate people within your organization about the risks associated with artificial intelligence and the various types of adversarial attacks. Emphasize the importance of cybersecurity awareness by encouraging proactive measures, regular software updates, security patches, and adherence to best practices in AI development and deployment.
- Education and Awareness: Encourage collaboration and information sharing within the AI community to spread knowledge about emerging adversarial threats and effective countermeasures.

Ethical Considerations

Establishing ethical guidelines for adversarial research and promoting responsible disclosure practices is crucial.

As stated by the White House, on October 20, 2023, President Biden issued an executive order to manage the risk of AI. In this order, AI systems with significant implications on national and economic security or public health and safety must undergo stringent safety testing and government supervision. With this executive order, the President directs the most sweeping actions ever to protect Americans from the potential risks of AI systems. Here are a few actions:

- **Developers of the most powerful AI systems are required to share their safety test results and other critical information with the US government.**

  In accordance with the Defense Production Act, the Order will require the companies developing any foundation model that poses serious risks to national security, national economic security, or national public health and safety to notify the federal government when training the model and to share the results of all red-team safety tests. These measures will ensure AI systems are safe, secure, and trustworthy before companies make them public.

- **Develop standards, tools, and tests to help ensure that AI systems are safe, secure, and trustworthy.**

  The National Institute of Standards and Technology will set rigorous standards for extensive red-team testing to ensure safety before public release. The Department of Homeland Security will apply those standards to critical infrastructure sectors and establish the AI Safety and Security Board. The Departments of Energy and Homeland Security will also address AI systems' threat to critical infrastructure, as well as chemical, biological, radiological, nuclear, and cybersecurity risks. Together, these are the most significant actions ever taken by any government to advance the field of AI safety.

- **Protect against the risks of using AI to engineer dangerous biological materials.**

  Develop strong standards for biological synthesis screening. Agencies that fund life science projects will establish these standards as a condition of federal funding. This creates powerful incentives to ensure appropriate screening and manage risks potentially made worse by AI.

- **Protect Americans from AI-enabled fraud and deception by establishing standards and best practices for detecting AI-generated content and authenticating official content.**

  The Department of Commerce will develop guidance for content authentication and watermarking to clearly label AI-generated content. Federal agencies will use these tools to make it easier for Americans to know that the communications they receive from the government are authentic – and set an example for the private sector and governments around the world.

- **Establish an advanced cybersecurity program to develop AI tools to find and fix vulnerabilities in critical software.**

  This builds on the Biden-Harris Administration's ongoing AI Cyber Challenge. Together, these efforts will harness AI's potentially game-changing cyber capabilities to make software and networks more secure.

- **Order the development of a National Security Memorandum that directs further actions on AI and security.**

  This is to be developed by the National Security Council and White House Chief of Staff. This document will ensure that the US military and intelligence community use AI safely, ethically, and effectively in their missions. It will direct actions to counter adversaries' military use of AI.

To better protect Americans' privacy, including the risks posed by AI, the President calls on Congress to pass bipartisan data privacy legislation to protect all Americans, especially kids. Below are a few of the actions directed:

- **Protect Americans' privacy by prioritizing federal support for accelerating the development and use of privacy-preserving techniques.**

  This includes ones that use cutting-edge AI and that let AI systems be trained while preserving the privacy of the training data.

- **Strengthen privacy-preserving research and technologies.**

  For example, cryptographic tools that preserve individuals' privacy, by funding a Research Coordination Network to advance rapid breakthroughs and development. The National Science Foundation will also work with this network to promote the adoption of leading-edge privacy-preserving technologies by federal agencies (The White House 2023).

  Similarly, the European Union created something similar called the EU AI Act, which are regulations on artificial intelligence.

  As stated in the High Level Summary of the Act provided by the Future of Life Institute, majority of the rules apply to AI systems that are considered high risk. High-risk systems are involved in the following areas: biometrics, critical infrastructure, educational and vocational training, employment, law enforcement, migration, administration of justice and democratic processes, and access to and enjoyment of essential private services and essential public services and benefits. In it, high-risk AI providers must

- Establish a risk management system through the high-risk AI systems' life cycle
- Conduct data governance, ensuring that training, validation, and testing datasets are relevant, sufficiently representative, and, to the best extent possible, free of errors and complete according to the intended purpose
- Design their high-risk AI systems to allow deployers to implement human oversight
- Design their high-risk AI system to achieve appropriate levels of accuracy, robustness, and cybersecurity (EU Artificial Intelligence Act)

This is only a few of the rules mentioned in this document. For this chapter, I focused on the rules that I felt could help with preventing adversaries and abuse within artificial intelligence. (EU Artificial Intelligence Act)

Other countries have implemented rules and considerations aside from those of the United States and the EU. China aims to be the leading artificial intelligence center by 2030. The Chinese Cybersecurity Law and the New Generation AI Development Plan outline data protection and cybersecurity strategies in AI, focusing on compliance and proactive risk management. Canada has created programs such as the Pan-Canadian AI Strategy and the Canadian AI Council to promote the responsible development of AI and tackle ethical concerns within the AI sector.

Conclusion

Adversarial attacks seriously threaten the reliability and security of machine learning models. As AI becomes integrated into our lives, the potential consequences of adversarial attacks become more severe. To mitigate these risks, a multifaceted approach is necessary. This involves robust defense mechanisms and continuous research. The potential consequences of these attacks make employing a multilayered defense strategy essential. Throughout this chapter, we have discussed several defense measures that can mitigate the risk created by adversaries and give you better safeguards for your models against adversaries. The importance of continuous research and improvement in security practices cannot be overstated - new attack methods are constantly evolving, requiring ongoing adaptation and innovation. Understanding adversaries and abuse is crucial for developing robust and secure systems. By studying the tools, techniques, and examples, we can better anticipate potential threats and develop mitigation strategies that create more resilient and trustworthy artificial intelligence systems.

CHAPTER 11

Working with Models

Introduction

Artificial intelligence (AI) is no longer a concept confined to the realms of science fiction; it has become an integral part of our daily lives and various industries. AI's presence is penetrating and ever-growing, from voice assistants like Siri and Alexa to advanced machine learning models driving innovation in healthcare and consumer products. However, alongside this rapid advancement, a spectrum of fear and misconceptions exists. Some worry about AI taking over jobs, invading privacy, or even surpassing human intelligence to a point of no return. This chapter aims to address these concerns, highlighting the reasons why we should embrace AI rather than fear it.

The Historical Context of Technological Advancement

Looking back at history, significant technological innovation has been met with a degree of skepticism and fear. The Industrial Revolution, for instance, sparked concerns about job displacement as machines began to replace manual labor. Similarly, the advent of the Internet brought about fears of privacy invasion, job loss, and data security. Yet, over

M. Attobrah, *Essential Data Analytics, Data Science, and AI*,
https://doi.org/10.1007/979-8-8688-1070-1_11

time, these technologies proved to be transformative, driving economic growth, enhancing quality of life, and creating new job opportunities that previously did not exist.

AI represents one of the next leaps in technological advancement, poised to bring substantial benefits. The rise of AI is sparking concerns about job displacement, rightfully so, especially in fields that involve routine, repetitive tasks. However, AI has the potential to generate new roles that did not exist before, for example, AI maintenance, ethics management, AI security engineer, AI art director, and prompt engineers. While certain jobs may be automated, others will shift to focus on human skills that are harder to replicate.

Understanding this historical context helps to allay fears and provides a perspective on how societies can adapt and thrive with new technologies. AI can be a transformative tool that can enhance productivity, create new roles, and improve quality of life – if managed with foresight, ethical oversight, and balance.

The Benefits of AI: Transformative Potential Across Industries

Healthcare: Improved Diagnostics and Personalized Medicine

AI has revolutionized healthcare by improving diagnostic accuracy and enabling personalized treatment plans. Machine learning algorithms can analyze medical images with higher precision than human radiologists, catching early signs of diseases like Alzheimer's. AI-driven predictive models assists doctors in tailoring treatments to individual patients, enhancing the effectiveness of medical interventions and reducing adverse effects.

Education: Personalized Learning

In education, AI facilitates personalized learning experiences for students. Intelligent tutoring systems adapt to individual learning styles and paces, providing customized resources and feedback. An example of this is Duolingo. Duolingo is an app used to learn different languages. It provides listening, speaking, and reading exercises at a pace and level according to the student's performance.

Manufacturing: Automation and Predictive Maintenance

AI enhances manufacturing processes by automating repetitive tasks and enabling predictive maintenance. Robotics powered by AI increase production efficiency and consistency, while predictive maintenance models anticipate equipment failures, reducing downtime and maintenance costs. Examples of these are Amazon's fleet of robots that work with the employees within their fulfillment centers to transport goods and IBM's Maximo Application Suite. This suite is used to do things like asset monitoring, predictive maintenance, and reliability planning.

Everyday Life: Personal Assistants and Smart Home Devices

In our everyday lives, AI manifests in the form of personal assistants like Siri, Alexa, and Google Assistant, making our lives more convenient by managing schedules, setting reminders, and controlling smart home devices. These advancements underscore AI's potential to improve daily living through automation and intelligent assistance.

AI Enhancing Human Capabilities

AI's true strength lies in its ability to augment human intelligence and decision-making. By handling repetitive and mundane tasks, AI frees up a person's capacity for creativity, strategic thinking, and problem-solving. For example, in business, AI can help identify patterns in markets and products, enabling more informed decision-making. In healthcare, AI can analyze and extract texts from documents to assist healthcare providers in making decisions faster. In law, AI can assist in finding relevant legal documents for their case in a sea of information.

Furthermore, AI enhances productivity across various sectors. In creative fields like design and music, AI tools can assist artists in generating new ideas and overcoming creative blocks. In research, AI can assist in accelerating discoveries by analyzing complex datasets and identifying patterns that lead to breakthroughs. By augmenting human capabilities, AI assists in fostering innovation and driving progress.

However, there are also downsides. Overreliance on AI for decision-making could lead to a reduction of critical thinking and problem-solving skills, as people may become too reliant on AI systems to perform tasks that require human judgment. This is particularly concerning in areas like healthcare or law, where human oversight is crucial to ensuring ethical decisions. Addressing these risks involves ensuring that AI serves as a tool to complement, rather than replace, human intelligence.

Successful AI Integration

Healthcare

Flo is an app focused on women's health. It tracks menstrual cycles, ovulation, symptoms, and more. The app uses a machine learning algorithm that leverages a person's previously recorded information, like age, period dates, and symptoms, to estimate cycle length predictions. It also offers personalized content like videos and articles to educate its end users. This content is reviewed by health and well-being experts to ensure the information in that personalized content is trustworthy.

Finance

Betterment is a financial investment platform that uses AI to create tailored profiles based on their financial plan. It uses algorithms to automate trading, transactions, and portfolio management. It was founded in 2008.

Consumer Goods

iRobot is a consumer robot company. They created an autonomous vacuum called the Roomba. It uses machine learning algorithms to create maps of the owner's property. This helps the vacuums avoid obstacles while cleaning. The end users can also label points within the generated maps in the app as "cleaning zones" to tell the Roomba to clean those specific points. For example, if your employees had lunch in office room HC, you could tell the Roomba to "clean under table HC."

Addressing Common Fears and Misconceptions

AI As a Threat to Privacy: Balancing Innovation with Ethical Considerations

Another common concern is that AI poses a threat to privacy. Data collection and analysis can lead to privacy issues if not managed properly. However, it is crucial to balance innovation with ethical considerations. Implementing robust data protection measures, cybersecurity requirements, and transparent data usage policies can mitigate privacy risks. Moreover, ongoing advancements in AI ethics, such as those mentioned in Chapter 10, aim to ensure that AI systems respect user privacy and operate within ethical boundaries.

The Myth of AI As an Uncontrollable Force

Ensuring proper oversight and regulation of AI development can help prevent unintended consequences over AI systems. The EU AI Act and US Executive Order on AI are frameworks and standards that are examples of regulations put in place to address AI-related risks.

Additionally, the AI Alliance – an international community of researchers, developers, and organizations – is actively working on building a community to drive responsible AI innovation. Their efforts emphasize maintaining scientific rigor while ensuring trust, safety, and diversity.

Responsible AI

Responsible AI is the approach of creating and deploying AI systems to ensure they align with ethical principles, legal requirements, and societal values.

Understanding Bias

Figures 11-1 to 11-6 shows some examples where AI systems have gotten it wrong ... very wrong.

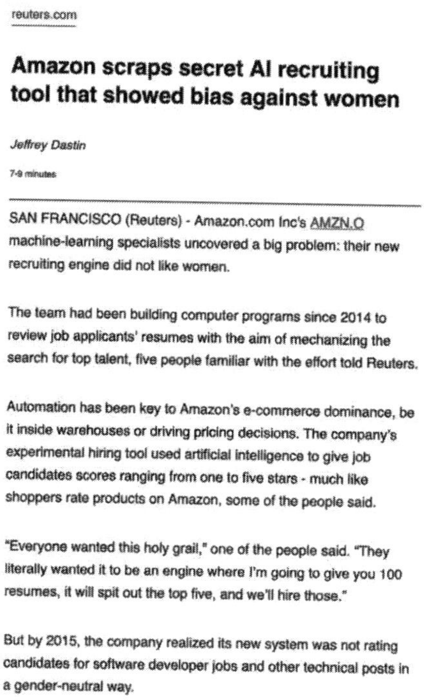

reuters.com

Amazon scraps secret AI recruiting tool that showed bias against women

Jeffrey Dastin

7-9 minutes

SAN FRANCISCO (Reuters) - Amazon.com Inc's AMZN.O machine-learning specialists uncovered a big problem: their new recruiting engine did not like women.

The team had been building computer programs since 2014 to review job applicants' resumes with the aim of mechanizing the search for top talent, five people familiar with the effort told Reuters.

Automation has been key to Amazon's e-commerce dominance, be it inside warehouses or driving pricing decisions. The company's experimental hiring tool used artificial intelligence to give job candidates scores ranging from one to five stars - much like shoppers rate products on Amazon, some of the people said.

"Everyone wanted this holy grail," one of the people said. "They literally wanted it to be an engine where I'm going to give you 100 resumes, it will spit out the top five, and we'll hire those."

But by 2015, the company realized its new system was not rating candidates for software developer jobs and other technical posts in a gender-neutral way.

Figure 11-1. *This image depicts Amazon's decision to scrap an AI recruiting tool due to bias. This highlights concerns over unfair treatment in hiring practices, which can be exacerbated*

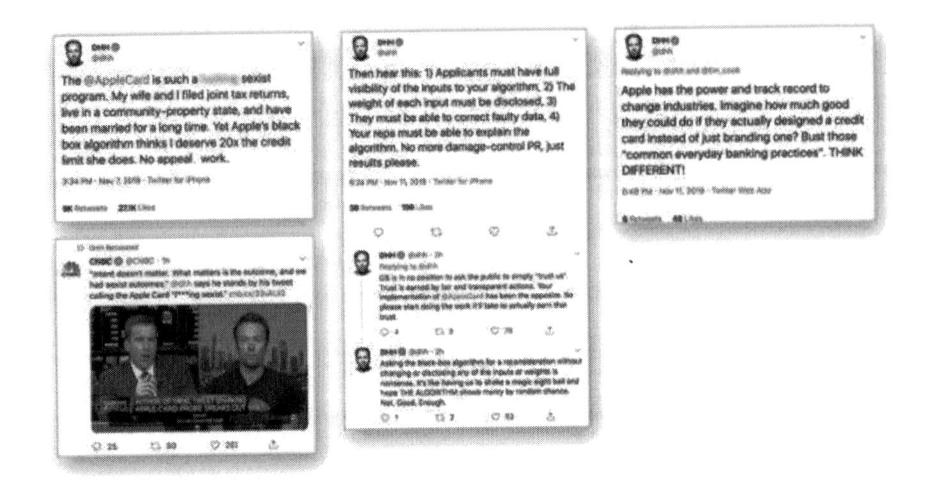

***Figure 11-2.** An image capturing David Heinemeier Hansson accusing Apple Card of gender discrimination, sparking a debate over biased credit limits based on gender*

***Figure 11-3.** An image illustrating how Twitter's automatic cropping of images did so in a biased manner. It consistently displayed white faces over black faces*

Figure 11-4. *An image showing Joy Buolamwini demonstrating the racial bias in AI systems. This facial recognition technology failed to see her face until she wore a white mask*

Figure 11-5. *An image showing the social injustice of a false facial recognition match leading to the wrongful imprisonment of an innocent Black man. This emphasizes the issues of racial bias in AI systems*

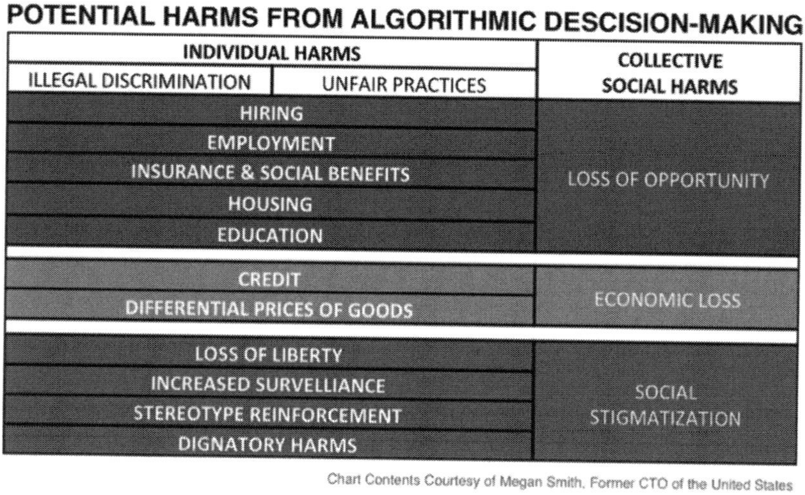

POTENTIAL HARMS FROM ALGORITHMIC DESCISION-MAKING

INDIVIDUAL HARMS		COLLECTIVE SOCIAL HARMS
ILLEGAL DISCRIMINATION	UNFAIR PRACTICES	
HIRING		LOSS OF OPPORTUNITY
EMPLOYMENT		
INSURANCE & SOCIAL BENEFITS		
HOUSING		
EDUCATION		
CREDIT		ECONOMIC LOSS
DIFFERENTIAL PRICES OF GOODS		
LOSS OF LIBERTY		SOCIAL STIGMATIZATION
INCREASED SURVELLIANCE		
STEREOTYPE REINFORCEMENT		
DIGNATORY HARMS		

Chart Contents Courtesy of Megan Smith, Former CTO of the United States

Figure 11-6. *An image highlighting the dangers of relying on algorithms, demonstrating how these systems can perpetuate loss of opportunities, economic loss, and social stigmatization. This can increase inequality and social harm*

Mitigating Bias – Human in the Loop Approach

Human in the Loop is a collaborative method that incorporates human expertise and input throughout the life cycle of the AI system. Humans play an active role in things like training, evaluating, and managing the system. This offers crucial guidance, feedback, data labeling, and transparency. The purpose of having humans in the loop is to increase the accuracy, reliability, and adaptability of AI systems by leveraging the strengths of both humans and machine learning models. Human involvement will be necessary to ensure decisions made from machine learning models are ethically sound as models become more complex and increasingly adopted by society. The goal is to avoid having AI systems make harmful decisions, as showcased in the "Understanding Bias" section above.

Ways Humans Can Be in the Loop but Not Limited To

- Providing data labeling and data quality and assurance before and after datasets are fed into the machine learning algorithm
- Providing feedback to machine learning models

Benefits of Humans in the Loop

- Enhanced Reliability
  - Human supervision is important for assisting in ensuring the models are reliable.
- Transparency
  - The human would have to have some understanding of each process that they are involved in within the machine learning process.
- Mitigation of Bias
  - Humans become part of the decision process. This helps identify and mitigate potential biases outputted by the model and in the training dataset.
- Employment Opportunities
  - This could bring new job opportunities for people, emphasizing that although this technology may take away some jobs, it could also bring forth new job opportunities for people.

Challenges of Humans in the Loop

It's important to look at all angles of a problem to understand the trade-offs before making a decision. One major challenge that arises in this process is speed. The process is no longer fully automated. However, the system outputs become more reliable. This process can reduce scale and increase expenses. However, your system reliability and trust from the clients/end users have the potential to increase. Depending on the tasks, another challenge is humans can also come with their own set of biases. However, one way to mitigate that is having a diverse set of people with an agreed-upon set of rules/requirements to consider when evaluating, providing feedback, and labeling datasets.

Mitigating Bias – Why Diversity Matters

Diversity in data science and machine learning teams is crucial for identifying and addressing potential biases. Along with having people with different educational backgrounds, individuals from various genders, ethnicities, and life experiences bring unique perspectives and observations, enabling the team to anticipate and mitigate challenges that might otherwise be overlooked in developing AI systems. This can save a company the harm to reputation and avoid potentially getting sued.

Building Trust and Transparency – Ethical AI: Ensuring Fairness, Accountability, and Transparency

The ethical deployment of AI is paramount to its acceptance and success. Ethical AI includes principles of fairness, accountability, and transparency. Fairness ensures that AI systems do not discriminate against individuals or groups, while accountability holds developers and organizations responsible for the outcomes of their AI systems. Transparency involves making the workings of AI systems understandable to users and stakeholders. This will foster trust and informed decision-making.

Several initiatives and frameworks have been developed to promote ethical AI. For instance, in Chapter 10, the Executive Order for AI was released in the United States by President Biden in 2023; the EU AI Act and other geographies all over the world provide a comprehensive set of requirements for developing AI systems. These guidelines emphasize technical robustness, privacy, and transparency, ensuring that AI systems benefit society while minimizing risk.

The Role of Education and Public Engagement

Educating the public about AI and its benefits is crucial for dispelling fears and fostering acceptance. Public engagement initiatives can demystify AI, addressing misconceptions and highlighting positive potential. Partnerships with community organizations, local governments, and non-profits can play a key role in reaching diverse populations, ensuring that everyone has access to accurate information and feels included in the conversation. Collaboration between technologists, policymakers, and the public is essential to ensure that AI development aligns with societal values and needs.

Transparent communication about AI advancements and their implications fosters trust and informed decision-making. Publicly available webinars, online courses, and AI literacy campaigns can further spread knowledge in accessible ways, allowing people to learn at their own pace. Engaging in open dialogue with stakeholders helps address concerns and ensures that AI systems are designed in ways that benefit society as a whole.

The Future AI: Opportunities and Challenges

The future of AI holds immense promise, with emerging trends and applications poised to revolutionize various fields. For example, AI-driven advancements in autonomous vehicles, natural language processing, and robotics are set to transform transportation, communication, retail, education, manufacturing, and other areas.

A great example of the future possibilities of AI, as of the writing of this book, is Apple Intelligence.

In June 2024, Apple announced Apple Intelligence, a suite of AI features the company plans to implement in its products. They plan to implement many features. I will mention a few. One of the features they introduced was the Math Notes calculator. One of the capabilities of this calculator would allow people to solve math equations as they write them down in the Math Notes application. Another feature of Apple Intelligence is Smart Script. This will assist in increasing the legibility of handwritten notes using the Apple Pencil by straightening text in real time. This could tremendously help those who have gotten the compliment that they write like doctors.

However, alongside these opportunities, there are challenges that need to be addressed. Ensuring the ethical use of AI, mitigating biases in AI models, and managing the societal impact of AI adoption are critical areas of focus. For example, the development of autonomous weapons raises ethical concerns. Additionally, the automation of jobs in industries has the potential to displace large segments of the workforce. By proactively addressing these challenges, we can harness the full potential of AI for the betterment of humanity. These include implementing ethical regulation, as discussed in Chapter 10, and preparing the workforce for new roles in an AI-driven economy.

Conclusion

In conclusion, AI represents a powerful tool with the potential to transform industries and enhance human capabilities. By embracing AI and addressing common fears and misconceptions, we can unlock its full potential while ensuring ethical and responsible development. The historical context of technological advancement, the tangible benefits of AI, and ongoing efforts to promote ethical AI provide a compelling case for embracing AI with a balanced perspective. As we move forward, it is essential to foster public engagement, educate stakeholders, and ensure transparency in AI development. By doing so, we can create a future where AI benefits all of humanity, driving progress and improving quality of life.

References

1. Hulten, G. (2018). *Building intelligent systems: A guide to machine learning engineering*. Apress
2. Eng. Emad Eldin Ibrahim Moselhy Bakr. (2024, April 23). *Ai-ml-DML-Gen ai- LLM sorting according to prices*. LinkedIn. `https://www.linkedin.com/pulse/ai-ml-dml-gen-ai-llm-sorting-according-prices-moselhy-ibrahim-bakr-7mz2f/`
3. Choosing the right estimator. *sci-kit*. (n.d.). `https://scikit-learn.org/1.3/tutorial/machine_learning_map/`
4. Cocheo, S. (April 6, 2022). *Tweetstorm blasts Goldman Sachs and Apple Card for "Sexist" Algorithm*. The Financial Brand. `https://thefinancialbrand.com/news/social-media-banking/apple-card-goldman-sachs-twitter-social-media-credit-discrimination-algorithm-90253/`
5. Mirjam Guesgen, Content Creator, @MirjamJG (n.d.). *Technical solutions to tackle AI bias*. Monadical Consulting. `https://monadical.com/posts/algorithm-bias.html#`

M. Attobrah, *Essential Data Analytics, Data Science, and AI*,
https://doi.org/10.1007/979-8-8688-1070-1

6. *BU Law Faculty and Students Take on Algorithmic Bias*. School of Law BU Law Faculty and Students Take on Algorithmic Bias Comments. (n.d.). https://www.bu.edu/law/record/articles/2019/bu-law-faculty-and-students-take-on-algorithmic-bias/

7. *Why this matters*. MIT Media Lab (n.d.). https://www.media.mit.edu/projects/gender-shades/why-this-matters/

8. Sarlin, J. (April 29, 2021). *A false facial recognition match sent this innocent black man to jail | CNN business*. CNN. https://www.cnn.com/2021/04/29/tech/nijeer-parks-facial-recognition-police-arrest/index.html

9. Insight - Amazon scraps secret AI recruiting tool that showed bias against women | reuters (n.d.). https://www.reuters.com/article/world/insight-amazon-scraps-secret-ai-recruiting-tool-that-showed-bias-against-women-idUSKCN1MK0AG/

10. Eng. Emad Eldin Ibrahim Moselhy Bakr (April 23, 2024). *AI-ML-DML-GEN AI- LLM sorting according to prices*. LinkedIn. https://www.linkedin.com/pulse/ai-ml-dml-gen-ai-llm-sorting-according-prices-moselhy-ibrahim-bakr-7mz2f/

11. Ackerman, E. (March 29, 2023). *Slight street sign modifications can completely fool machine learning algorithms*. IEEE Spectrum. https://spectrum.ieee.org/slight-street-sign-modifications-can-fool-machine-learning-algorithms

12. The United States Government (October 30, 2023). *Fact sheet: President Biden issues executive order on safe, secure, and trustworthy artificial intelligence.* The White House. https://www.whitehouse.gov/briefing-room/statements-releases/2023/10/30/fact-sheet-president-biden-issues-executive-order-on-safe-secure-and-trustworthy-artificial-intelligence/

13. *Annex III: High-risk AI systems referred to in Article 6(2).* EU Artificial Intelligence Act (n.d.). https://artificialintelligenceact.eu/annex/3/

14. Savio Jacob (April 30, 2024). *AI regulations around the world - spiceworks.* Spiceworks Inc. https://www.spiceworks.com/tech/artificial-intelligence/articles/ai-regulations-around-the-world/amp/

Index

M. Attobrah, *Essential Data Analytics, Data Science, and AI*,
https://doi.org/10.1007/979-8-8688-1070-1

D

F

G

H

I

J

K

L

M

N

O

S

T

U

V

W, X

Y, Z

Printed in the United States
by Baker & Taylor Publisher Services